QUENCH THE FIRE

A SEARCH FOR BALANCE IN THE UNDERSTANDING OF CLIMATE CHANGE

TERRY LUCAS

Quench the Fire
Copyright © 2021 by Terry Lucas

All rights reserved. No part of this publication may be reproduced, distributed, or transmitted in any form or by any means, including photocopying, recording, or other electronic or mechanical methods, without the prior written permission of the author, except in the case of brief quotations embodied in critical reviews and certain other non-commercial uses permitted by copyright law.

Tellwell Talent
www.tellwell.ca

ISBN
978-0-2288-5892-8 (Hardcover)
978-0-2288-5891-1 (Paperback)
978-0-2288-5893-5 (eBook)

This book is dedicated to:
My siblings. Those who have passed, and those who remain.

TABLE OF CONTENTS

INTRODUCTION

THIS BOOK IS NOT A record of all that I know about the subject of climate change at the time of writing. Rather it is a journey of discovery for me, where I start here with some basic knowledge and interest and gradually, chapter by chapter, reveal to myself and to the reader whatever I find to be rational and well supported by evidence.

That is not to say that I have no plan. I have arranged 7 chapters which address one by one the principles that need exploration in order to arrive at the end with a possible plan that I would humbly recommend to decision-makers who might see fit to consider.

In the first chapter I will explore the evidence that climate change exists, and that it is caused by human activity. For now, I am assuming that I will find these two basic tenets to be true, although obviously I must keep an open mind. In the second Chapter I will look at what effects, if any, that climate change is having in today's world. In the third, I will look at projections of what might be expected in the future if climate change continues at rates predicted by the experts. In the fourth, I will explore the history and impacts of fossil fuels. In the fifth I will evaluate the solutions to emissions reductions that are currently active in today's economies. In the sixth I will attempt to tackle the beast that I call "The Geopolitical Landscape". Finally, armed with the sum total of my new knowledge, I will attempt in the seventh Chapter to explore future solutions and make some recommendations.

Why am I doing this? Firstly, I am a scientist myself, although I have no particular skills in many of the various sub-disciplines that make up climate change science. I am, however, true to the principles of scientific endeavor. My natural curiosity has in the past led me to dig deeper into claims made about climate science that I find in the media, and often I have found faulty science, or no science at all. A large collection of such experiences leads me here. As you read on, you will find real world examples of what I mean.

My background is in Mechanical Engineering, and in the course of my career I have come to be something of an expert in a few sub-disciplines, such as heat transfer, assessment of structural load capability, and several subjects associated with the design and operation of gas turbine engines. As I gained experience, I was tasked with integration of many sub-disciplines, including some with which I was only superficially familiar, to arrive at solutions to multi-disciplinary problems. This makes me somewhat qualified to do the job I am taking on here. It is important to understand, however, that there are myriad disciplines and sub-disciplines involved, making no-one on Earth a true expert in this broad subject.

I know in advance that I will be doing small calculations that will serve to validate or invalidate assumptions being made about the science. You will find these are simple calculations and I will undoubtedly have to explain why they still work to prove the points I am making. It is true that extraordinarily complex phenomena require in turn complex calculation methodologies to characterize them, but it is also true that simple, fundamental principles cannot be violated.

Let me give an example right here so that my approach is clear and well understood. If a person wishes to climb a mountain, taking the route of least resistance and requiring the minimal energy expenditure, she may have to look at 3D topographical maps of the various ascent routes to find the optimal one. Longer routes would be less steep, but energy is expended even over flat terrain, so why assume that the long ones are the easiest? Perhaps the shortest, steepest route is actually the best one. Or maybe a middle ground is best. Calculations can be done using her previously characterized metabolism at different paces and different levels of physical stress to arrive at the best solution.

Such calculations require a lot of input data and a certain level of complexity in the analytical process. On the other hand, there is one fundamental principle that cannot be avoided: the absolute minimum possible energy expenditure has to be equal to her body weight multiplied by the change in elevation. This is indeed a simple calculation to do, and it is mostly at that level that I will be operating as I take my journey.

For readers with less interest in the detailed calculations, the various charts and conclusions drawn from them can be accepted at face value. In this case, the book should read fluidly. For those requiring validation of the conclusions, the calculations and assumptions will be well described and subject to challenge.

Given the book will likely be filled with calculations and charts, the following question comes to mind, and only because I have imagined it being asked by a prospective publisher: Who is the target audience for this book?

Here is a list of possibilities:

> Members of the Climate Change science community who are steeped in their sub-disciplines but might like to be more informed about the entire scope of the issue.
>
> A sector of the general public that wishes to get more informed about Climate Change, but not so much that they want to read all of the scholarly papers.
>
> Policy makers in governments and private enterprises who would like to have in one place a simple guide with enough validated facts to make them conversant on the issues.
>
> The usual groups of climate science deniers and climate change alarmists who may try to extract some unintended ammunition for their causes.

Finally, I would like to describe here one of my hopes for an outcome of this book. I have called it "Quench the Fire", and of course I mean that literally and figuratively. The literal one is created by the burning of fossil fuels that reportedly (pending Chapter 1!) is contributing to anthropogenic Climate Change. The figurative ones are the fires created by overheated dialogues on all facets of the issue. I would like to hear respectful discussions that engage people in committing to common goals, and I would like to feel sure that those discussions are firmly grounded with factual information that everyone can support. That is too lofty a goal for this book alone, but I am hoping it can be a contributor.

CHAPTER 1

THE GREENHOUSE EFFECT

INFORMATION ON THE BASICS OF how warming is thought to occur is readily available on the internet. Typically, it is limited to overviews of the problem which by themselves are insufficient to be convincing from a scientific point of view. It basically says: "Trust me, the scientists all agree". On the other hand, it is possible to find learned papers on the subject which can be quite daunting for the uninitiated (which certainly includes me).

I have managed to extract the science to the best of my ability. I hope it is interesting and informative.

Ultimately, the Earth's only source of heat is our sun, and the Earth also emits heat to its surroundings. In this case, the surroundings are the atmosphere, and outer space.

Heat is transferred by conduction, convection, and radiation. Conduction is heat transferring inside a body from one boundary to another. If you add heat to one end of a metal bar, you will soon feel heat arrive at the other end. Convection is heat transferred from a fluid to a solid body at the interface by friction that occurs as a result of fluid motion. You can feel heat being drawn away from your body by a cold wind, and the amount is certainly a function of the wind speed. Finally, radiation is heat carried by electromagnetic waves, travelling to and from any body interacting with its surroundings. On a balmy 20°C day, it is comfortable in the shade, but it can feel a lot hotter in the sun, which radiates heat directly to our bodies.

All of these mechanisms of heat exchange are at work on the earth, at all times. Models that attempt to predict temperatures of the atmosphere and the ocean waters are very sophisticated and need a great deal of calibration from terrestrial measurements. However, it is observed that there are self-regulating mechanisms that maintain average temperatures relatively constant, and important changes are typically only observed over centuries or millennia. Presumably, this is because, in the end, there is a relatively fixed amount of heat coming from the sun, when averaged over the yearly cycle. Despite the many interactions that cause heat exchange, on average the Earth has arrived at an equilibrium.

The image below, taken from the US National Weather Service, shows schematically many of the processes at work. From the point of view of maintaining heat balance, radiation is the dominating mechanism. Convection and conduction are all terrestrial phenomena in a closed loop.

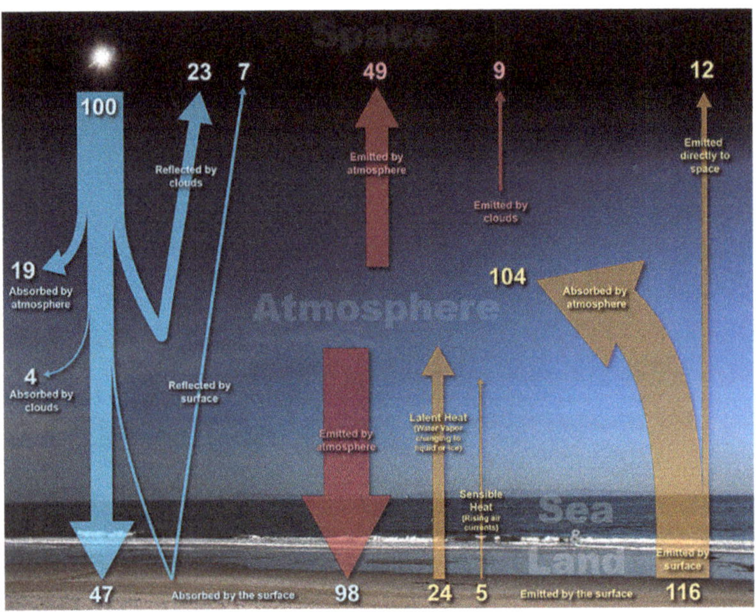

It is important to understand this summary of heat exchange before proceeding. The absolute value of the individual numbers may be questioned, but roughly this provides a good overview to start our exploration of climate change science. The figure is divided into 3 regions: on the left is the heat exchange with the sun, in the middle is heat exchange with the atmosphere, and on the right is heat exchange with the earth. All components are related to each other in some way. About half of the sun's energy is absorbed by the earth. The rest is absorbed by the atmosphere or reflected away. Most of the heat emitted by the earth is absorbed by the atmosphere. Then, the atmosphere radiates back to earth and also to outer space. These processes have reached an equilibrium over many millions of years, meaning that the mean temperatures of the earth's surface and the atmosphere are stable.

Exchange between the Earth and the atmosphere is the primary mechanism that we need to discuss here. In the schematic below, it is shown that the temperature of the radiating body determines the wavelengths at which radiation is emitted. The sun has roughly a surface temperature of 5800°K. It consequently emits radiation at short wavelengths. The temperature of the Earth's surface averages out to 290°K (17°C), so it emits radiation at much longer wavelengths. See image below from the American Chemical Society, with yellow representing the sun and red the Earth.

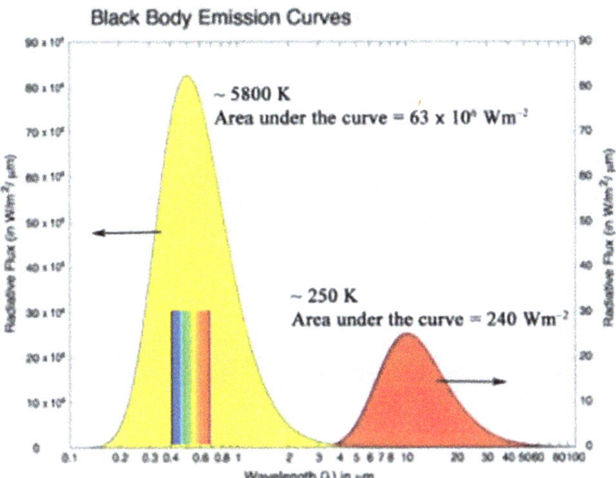

If the atmosphere were nonexistent, the earth would release much more heat to space than it currently does. Consequently, it would be much colder, more like the moon. Our atmosphere is in fact one giant greenhouse that keeps some of the sun's heat trapped, much to the benefit of all life on Earth.

The chemical composition of the atmosphere determines its propensity to either capture heat or allow it to escape to space. Each molecule has its own signature. The primary components of dry air are nitrogen (N_2) at 78%, Oxygen (O_2) at 20.9%), and Argon (Ar) at 0.9%. That leaves only 0.2% for all the remaining gases, including carbon dioxide (CO_2) at around 0.04%. Another important content is water vapor which varies but can reach 2%.

The figure below from Wikipedia combines the effects of incoming and outgoing radiation, including the effects of individual gases.

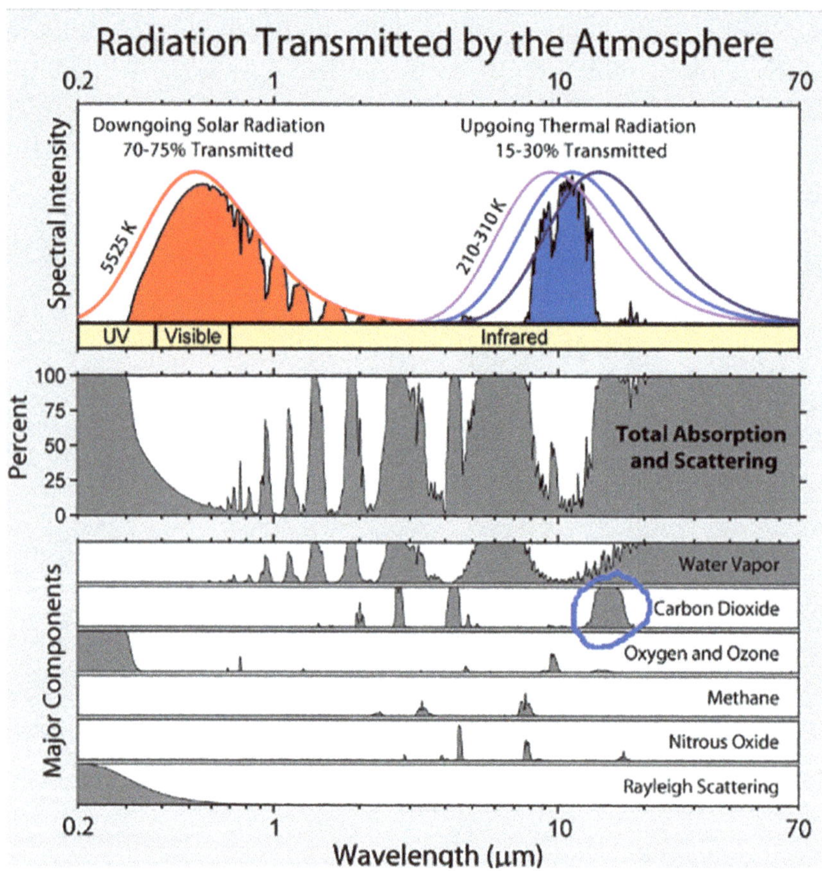

The data behind this graph is simply the well-known measured radiative scattering capabilities of different gases, across the whole wavelength spectrum. These have been measured in laboratory experiments.

In the UV (Ultraviolet) regime, oxygen and ozone play the important role of blocking potentially harmful radiation.

Water vapor plays a significant role, both in absorbing and scattering both incoming short-wave radiation and outgoing long-wave radiation. It is unquestionably the most important greenhouse gas.

CO_2, Methane and Nitrous Oxide are shown to have little or no effect in blocking incoming radiation, but they have a role to play in absorbing radiation from the Earth's surface, and thus are considered greenhouse gases.

Nitrogen is missing from the figure, even though it is the most abundant molecule in our atmosphere. It allows 100% transmission of radiation in all frequency ranges, and therefore plays no role in radiative heating or cooling. It does, however, act to absorb heat generated by greenhouse gases and therefore provides a warm blanket around our planet.

Although methane (CH_4) and Nitrous Oxide (N_2O) are much more powerful greenhouse gases than CO_2 for a given quantity of gas, they exist in much smaller quantities. In the table below, taken from Wikipedia and derived from numerous sources, the net relative effect of these gases can be seen in the last column. CO_2 is by far the largest contributor, but the other 2 are still significant.

Gas	Pre-1750 Concentration	Recent Concentration	Absolute Increase	Percentage Increase	Increased Radiative Forcing (W/m2)
Carbon Dioxide CO_2	280 ppm	395 ppm	115 ppm	41	1.88
Methane CH_4	700 ppb	1762 ppb	1062 ppb	152	0.49
Nitrous Oxide N_2O	270 ppb	324 ppb	54 ppb	20	0.17

The role of water vapor cannot be underestimated. If the presence of other gases like CO_2 causes warming, then the air is able to hold more water vapor. Expressed differently, the relative humidity might stay the same, but the absolute value of water vapor will increase. Since water vapor is a greenhouse gas, it will exacerbate any warming effects caused by the other gases, causing escalation of the warming effects.

We have seen, however, that water vapor also blocks incoming heat from the sun, in the wavelength range of .8 to 1.5 μm (microns, or micro-meters). So, wouldn't additional water actually cool the planet? The answer is no, and the reason can be seen in the first figure above. The amount of radiation from the sun that is absorbed by the atmosphere is only 19%, while radiation from the earth to the atmosphere is 104%. The latter is by far the most powerful effect.

Now, let us summarize where we have gotten so far. The Earth receives heat from the sun, and to maintain equilibrium, some of it is radiated back out into space. There are a few gases in the atmosphere that are reported to block outgoing radiation, thus trapping heat. These gases exist only in trace quantities, yet they are reported to have powerful effects.

Next, we should look at measurements that have been taken to validate greenhouse effects.

EVIDENCE OF GREENHOUSE GAS IMPACT ON OUR PLANET

The first thing to consider is how the composition of the atmosphere has been changing. We saw in the previous section that CO_2, CH_4 and N_2O can be important heat-trapping gases, because they scatter radiation specifically at the wavelengths being emitted by the planet. Below are graphs showing how the concentrations of these gases have been evolving: (Bernhard Bereiter, 2015). Included are concentrations of CFC's and HFC's which I will not be considering in this book as they have small effects.

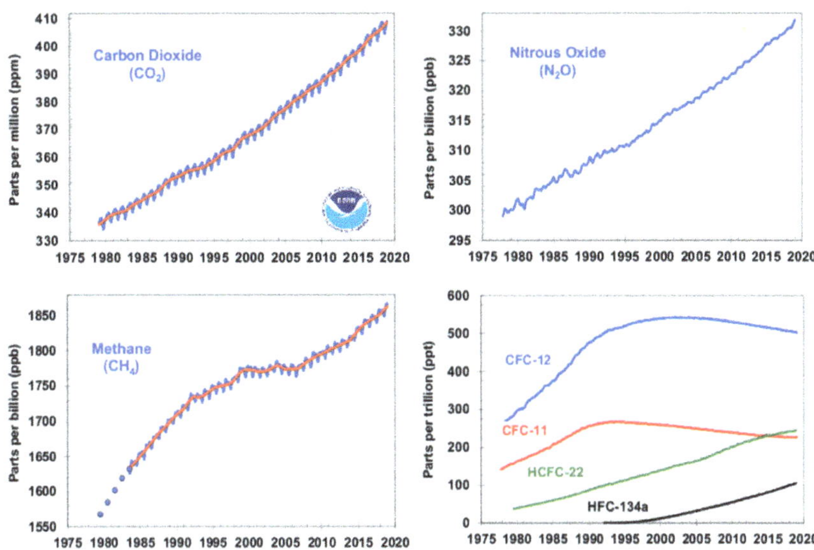

In the time scale shown on the graphs, technology required to measure these concentrations has been readily available and extremely accurate. The only calculation in the data is the necessity to average over large areas and times to catch the trend. For example, CO_2 emissions have small yearly fluctuations caused by natural processes related to seasonal growth and decay of plant matter.

In order to look at ancient times and relate global temperatures on a longer scale, scientists have developed a process where air bubbles contained in glacial ice can be extracted and their composition measured. Certain glaciers continue to grow year after year, so drilling out a "core" to a great depth can produce a long cylindrical sample which contains gas content history covering millennia. For CO_2, the graph below shows the trend over an 800,000-year scale! (Bernhard Bereiter, 2015)

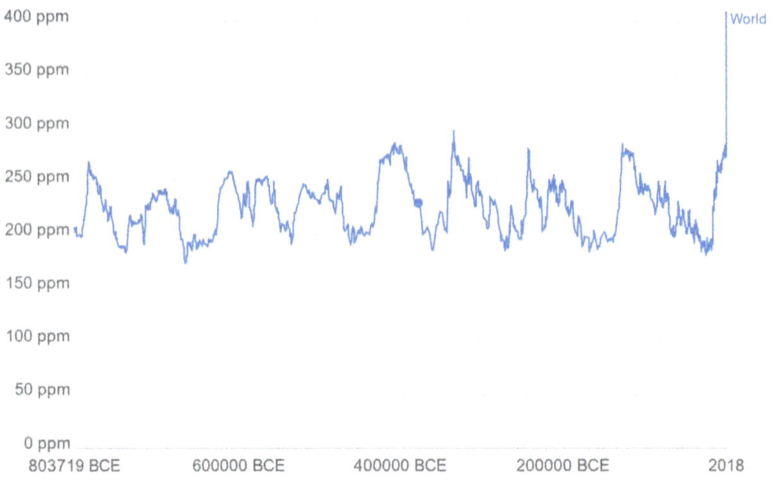

The sudden rising peak on this scale certainly coincides with industrialization, intricately linked with the burning of fossil fuels.

There is another way to check if the increase is caused by fossil fuels, or whether it is part of some as-yet-undiscovered process. All CO_2 contains small amounts of radioactive Carbon 14, which has a half-life of 5730 years. Any CO_2 newly created by other sources will have a certain known amount of Carbon 14 in it. But the content of Carbon 14 in the atmosphere's CO_2 has been decreasing. Below is one figure with data recorded in Norway. (Lövseth, 1996)

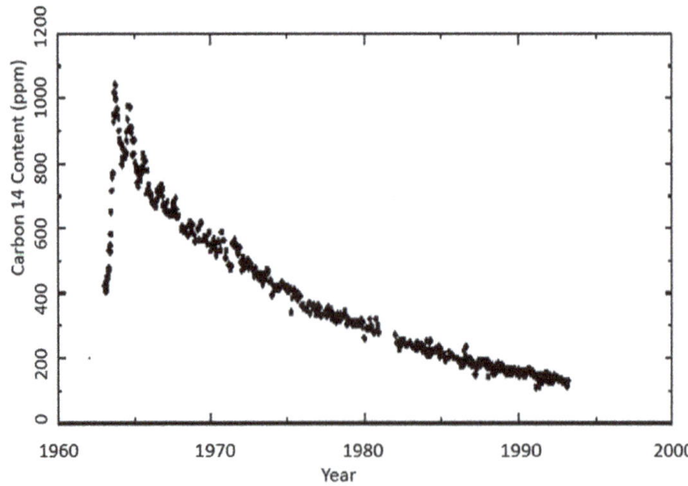

The spike seen in the 1960's was caused by nuclear bomb testing, which produces Carbon 14 in the atmosphere. Since then, the content has decreased, and at a rate much faster than normal decay of the isotope.

Fossil fuels are, by definition, incredibly old. The processes required to produce them from decayed organic matter are slow going, taking typically millions of years. Accordingly, they are almost entirely depleted of Carbon 14. Adding CO_2 produced by the combustion of fossil fuels will then naturally deplete the Carbon 14 content in the atmosphere, when expressed as a percentage of the total. This is just what is happening.

Regarding Methane CH_4, below is a chart showing human sources of methane. (P. Bousquet, 2006)

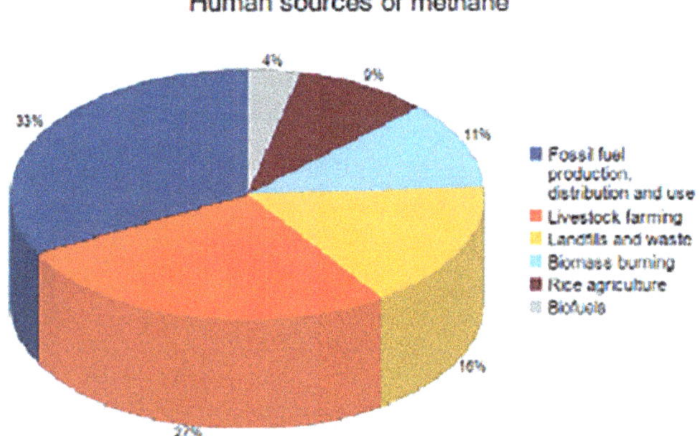

The production of fossil fuels creates methane emissions because it is a gas naturally trapped in the earth. It is released when the coal, oil or natural gas is extracted. In fact, natural gas is mostly methane and leakage into the atmosphere is unavoidable.

The other major source is livestock raised for meat production. The digestive tracts of ruminant farm animals, such as cows, sheep, and goats, naturally produce methane. It is released in their exhalations or in flatus.

Nitrous Oxide N_2O comes mainly from Nitrogen fertilization in agriculture, see below. (Arti Bhatia, 2012)

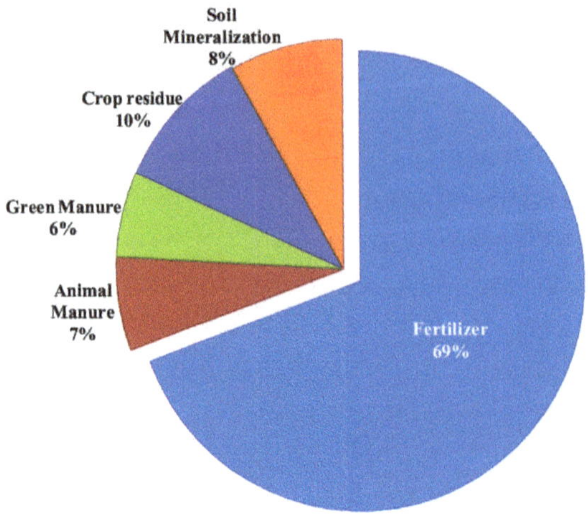

The increases in the presence of these gases in our atmosphere are almost indisputably a function of human activity, specifically the extraction and burning of fossil fuels, and agriculture. The rates of increase of the gases correspond well to increases in fuel consumption since the start of the industrial age, and to population growth which in turn drives more fuel consumption and more food production. Below is a graph showing relationship between population and CO_2 emissions, with raw data taken from Our World in Data.

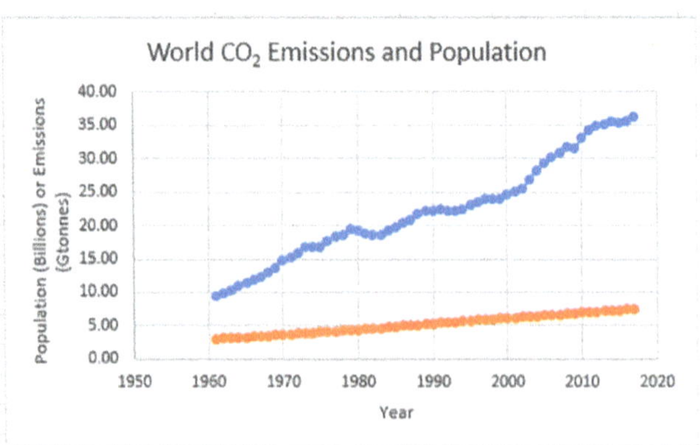

Since 1961 annual emissions have gone up by 282% while population has increased at only half of that rate, 144%. Therefore, we have been increasing our consumption over time even on a per capita basis. We will see later that the emissions growth tracks Gross Domestic Product (GDP) more closely than population.

It is important to consider what happens to CO_2 after it enters the atmosphere. For it to be depleted, certain chemical reactions must take place. The most important is the capture of carbon in plant life which results from photosynthesis. It is no secret that the Earth's landscape has seen a steady depletion of green spaces, especially forests. We cannot expect that CO_2 can be removed from the air as quickly as it has in the past unless we can reverse the trend of deforestation. Worse, the primary reason for deforestation has been the creation of arable land to feed the world's growing population. As we have seen, agriculture produces other greenhouse gases which can only exacerbate the effects.

Another source of CO_2 depletion is ocean water. The oceans absorb a lot of CO_2. Some gets dissolved, remaining as CO_2 in solution. Some gets used by algae in photosynthesis. Finally, some gets converted to carbonic acid which is contributing to a steady acidification of the oceans, causing potentially nasty effect on shellfish and the food chain. We will explore this last part in Chapter 3.

To summarise, here is a chart showing the distribution of greenhouse gases and their sources in the year 2010. To compare one to the other, they have all been converted to CO_2 equivalent in terms of their greenhouse effects. (Victor, 2014)

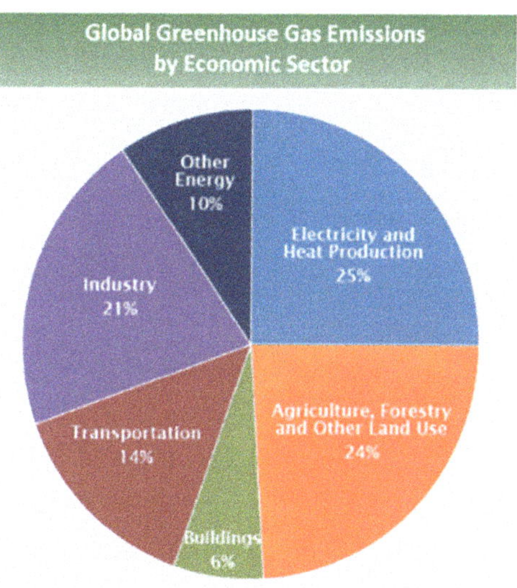

We see that Agriculture is a big contributor, even more important than Transport, which gets most of the press. For clarification, "Other Energy" is used by Energy Industries in the management of fuels. It includes energy needed to extract and transport fuels, as well as "Flaring" and Fugitive Gases". Flaring is the deliberate combustion of leaking fuels, mostly natural gas, to prevent it from entering the environment. Fugitive gases are those that leak without being combusted.

One further chart shows the evolution of the emission of greenhouse gases.

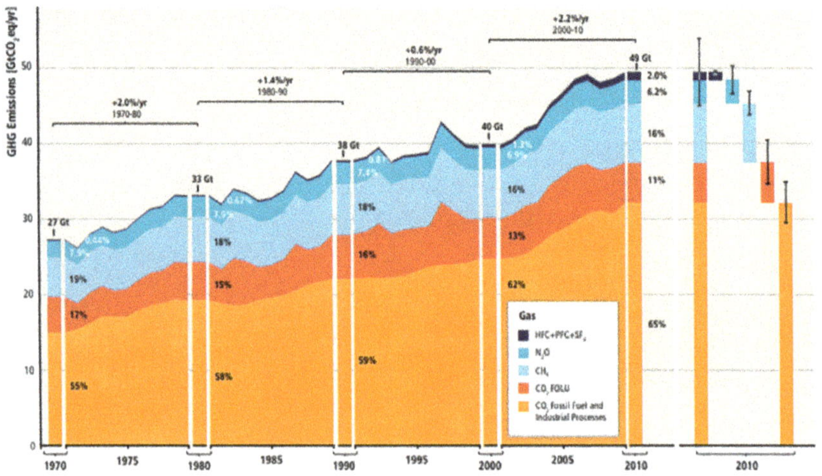

It is noteworthy that CO_2 from fossil fuel combustion accounts for only 65% of the total greenhouse effect of human–driven climate change.

EVIDENCE OF THE WARMING OF OUR PLANET

There are 4 sources of data for measuring temperatures with sufficient accuracy and coverage to show global trends. NASA GISTEMP is the most comprehensive, with 6300 measurement points covering 99% of the planet. Measurement points are naturally more abundant on land than over the oceans. Scientists use averaging and interpolations between measurement points to get a picture covering the whole globe. GISTEMP has a calculation point at every 2° latitude by 2° longitude area.

Below is a graph showing the temperature trends containing all 4 datasets. The 2 additional curves identified as "Berkeley Earth" and "Cowtan and Way" are re-calculations by different researchers using the same base data. Although some variations exist, the trend is very consistent, showing warming of about 1.25°C since the pre-industrial era. (Zyss, 2019)

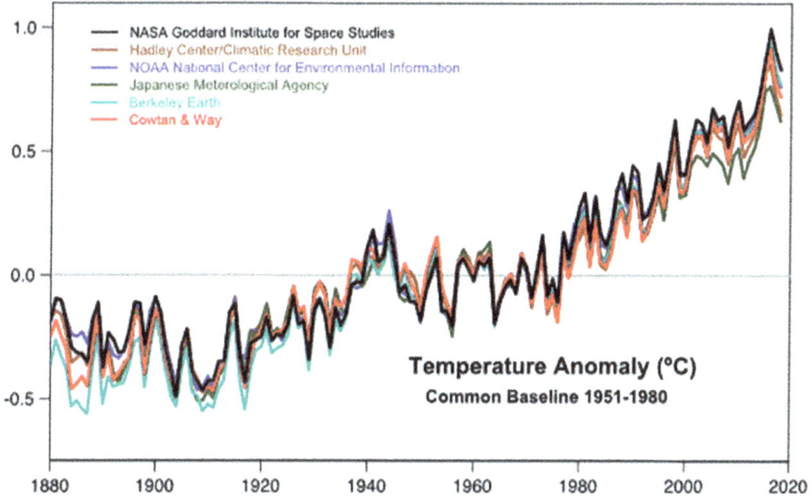

The same data is used to show visually the trends by region. In the figure below, the top image uses the HadCRUD data and leaves area blank wherever there is thought to be insufficient data, so this image could be considered "raw" data. Below that, Cowtan and Way fill in the gaps by extrapolating the data. They find higher temperature increases in the Arctic.

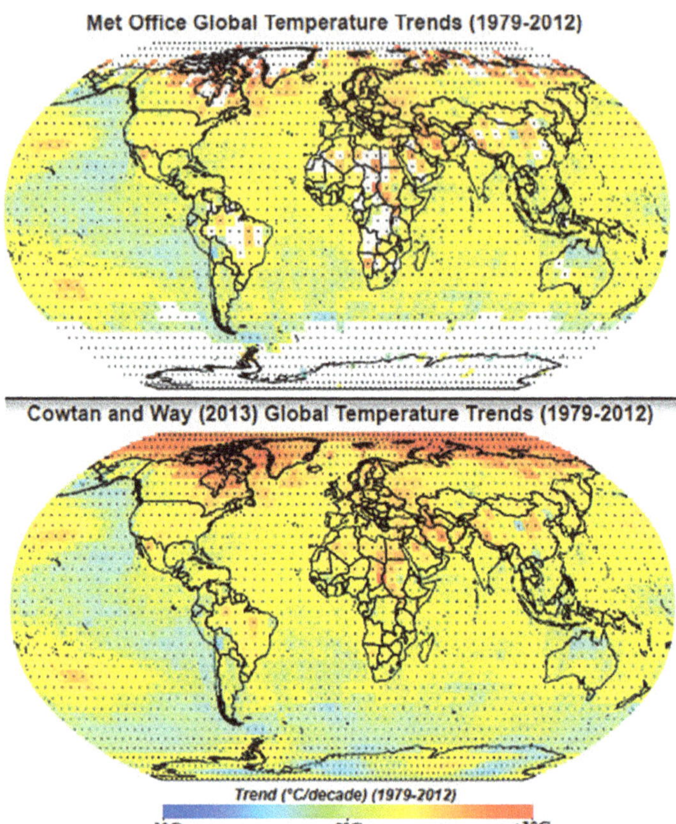

The origin of this data is a set of thermocouples measuring air temperature. Another way to validate it is to look at a completely different source and means of measurement. Satellites in space take infrared images of the earth's surface on a regular basis, and there is a historical trend available. Below is a figure produced by infrared imagery. These measurements are for the lower tropospheric part of the atmosphere, which is the part closest to the earth's surface. (Zyss, 2019)

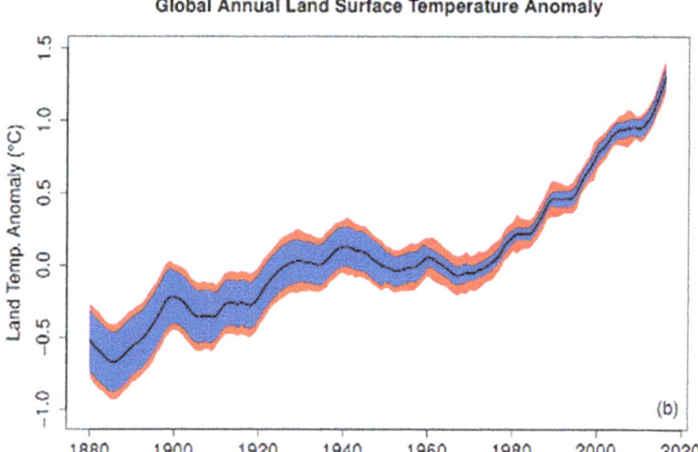

Global Annual Land Surface Temperature Anomaly

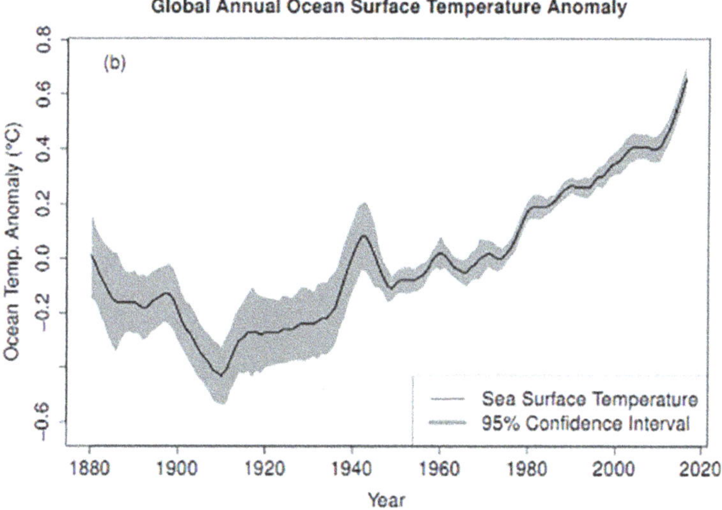

Global Annual Ocean Surface Temperature Anomaly

In these figures the width of the line is the uncertainty band, to 95% confidence level. Uncertainty is calculated based on known or inferred inaccuracies that could exist in the measurement and averaging processes. Since the uncertainty bands are themselves smaller than the average temperature change, it is concluded that warming is happening to a high degree of certainty.

Below are the combined land and ocean temperature measurements from the satellite imagery.

Annual Mean Temperature Change for Land and for Ocean

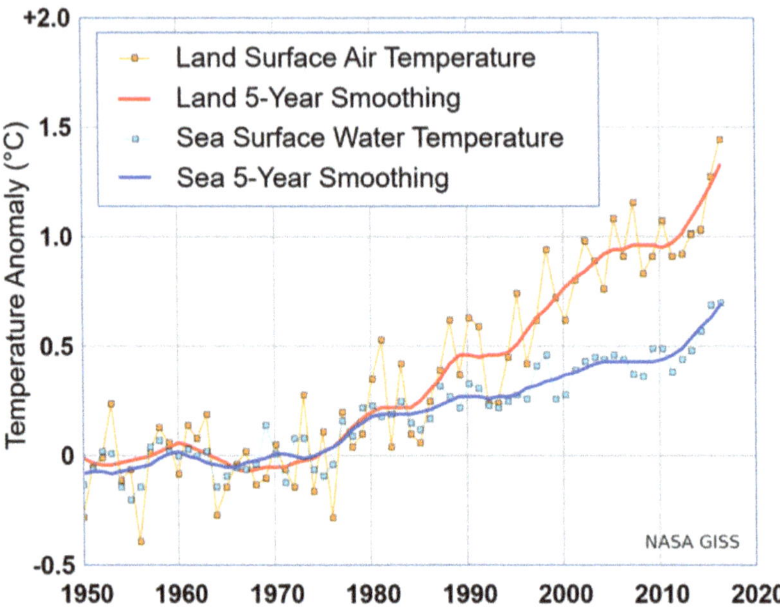

Measurement of ocean water temperatures is perhaps more important that measurements of air temperatures. Radiation from the atmosphere, and directly from the sun, are the largest contributors of heat to the oceans. Ocean water covers ⅔ of the earth, and it is a near-perfect black body, meaning it readily absorbs heat from radiative sources. It is estimated that 93% of the heat produced by global warming should be absorbed by the oceans. Water also has a heat capacity 4.2 Joules/Gram °C, which is the highest of any common substance. By contrast the heat capacity of air is 1 J/g°C. This means that it takes 4.2 times the amount of heat added to water to cause its temperature to rise a given amount, as compared to air. A rise in ocean temperature over a long time period is therefore a strong indication of a warming trend.

There is a much smaller distribution of physical measurement stations in the ocean water, as compared to those measuring air temperature. There are about 3000 "drifters" out there, measuring water temperatures at different depths. There is obviously a temperature gradient, from warmest at the surface to colder at greater depths. This data is used to calibrate the more comprehensive measurement system using Infrared radiation measurements from satellites, which measure radiation from a very thin layer on the surface.

Below is another graph that I find the most dramatic because it covers a 2000-year period and shows a sharp sudden increase in only very recent times. This does not appear to be part of a natural cycle. (Gerald R. North, 2006, p. 1)

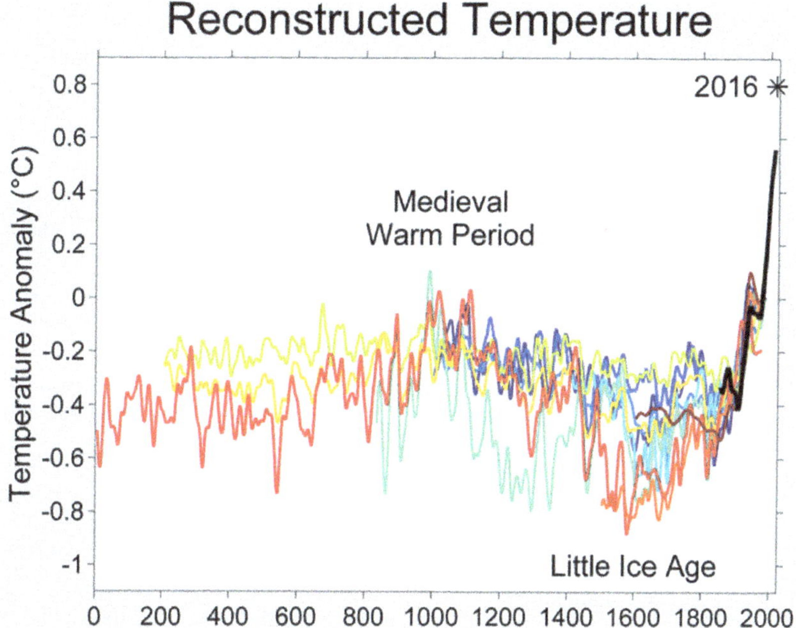

^ National Research Council (U.S.). Committee on Surface Temperature Reconstructions for the Last 2,000 Years *Surface temperature reconstructions for the last 2,000 years* (2006), National Academies Press ISBN 978-0-309-10225-4

The data going back to 1880, as shown on previous graphs, are quite reliable because they are obtained by direct measurements with reliable instrumentation. It is evident that such instruments were not available in ancient times. The data prior to 1880 is constructed using indirect measurements, such as: tree ring growth, coral growth, ocean sediment analysis of certain isotopes of oxygen, and data contained in ice cores of glaciers. The reason for the presence of multiple datasets is that each is a study from a different researcher, using mostly the same data. The variations from one dataset to another can then be considered the confidence band. Post–1880, the datasets all come together because, as stated previously, the measurements are consistent and reliable.

WHAT ARE THE SKEPTICS SAYING?

HOW CAN TRACE AMOUNTS OF A GAS CAUSE SUCH ENORMOUS EFFECTS?

CO_2 in the air has risen from .03% to .04% in about 75 years. That is indeed considered a trace amount of gas. The most scientific explanation is, of course, that the amounts of CO_2 and other greenhouse gases can be calculated to be having roughly the effects we see, based on the well-understood properties of these gases.

But there are certainly many examples in nature of small quantities having large effects. Consider that most medications we take are in the range of 200 to 1000 milligrams, while our body weight is in the range of 100,000 milligrams. If you remove water, protein, and fat from our bodies, we are left with only 6% of our total mass making up all of the other minerals needed that make us human.

Consider that the entire range of temperatures existing in the universe is from -273°C, absolute zero, to many millions of degrees inside of stars. In that case we might say that a 1°C rise is in fact a "trace" amount of temperature, in line with the trace amounts of gas causing the effect. But it so happens that the amount of 1°C is important indeed, considering that all life as we know it must exist between 0°C (freezing point) and 100°C (boiling point). Life on earth is adapted to a very narrow range of temperature, so a change of 1°C needs to be considered important.

IF CO_2 HOLDS HEAT IN, WHY DOESN'T IT ALSO KEEP HEAT OUT?

Going back to the first figures in this chapter, we can see that CO_2, CH_4 and N_2O have no filtering effect for the sun's incoming radiation because it is emitted at short wavelengths. It so happens that radiation from the Earth, called long-wave radiation, is in exactly the regime where these gases can have a reflective or scattering effect.

WATER VAPOR IS BY FAR THE MOST POWERFUL GREENHOUSE GAS AND IT ACTS IN THE SAME WAVELENGTH RANGE AS CO_2. SURELY THE EFFECTS OF CO_2 ARE NEGLIGIBLE IN THIS CASE.

Firstly, CO_2 has its own long wave radiative signature, similar to water vapor but not entirely overlapping. Water vapor will always exist in relatively constant amounts due to the mechanisms of the water cycle. But CO_2 is being added continuously, and mechanisms to diminish it are insufficient.

The existence of water vapor is also very much contingent upon air temperature. At higher altitudes, the air is less capable of holding water vapor, which is why clouds form.

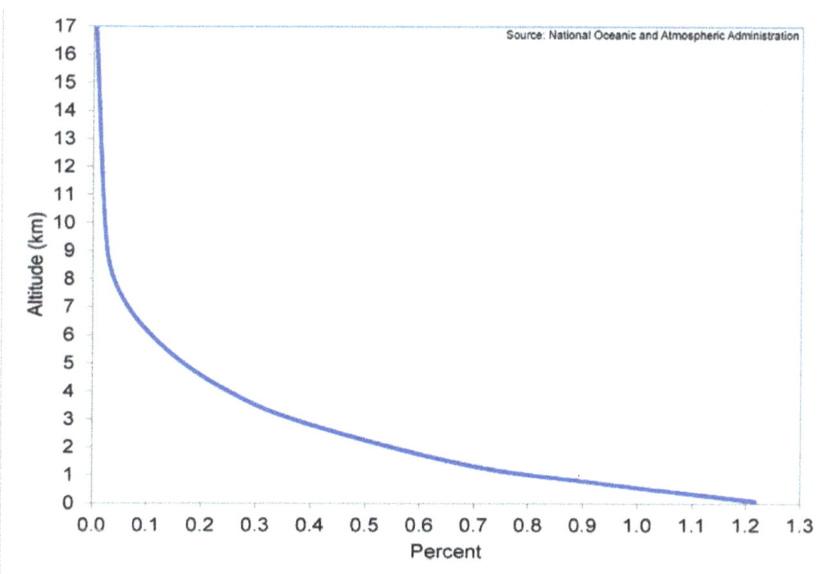

Change in Water Vapor Concentration with a Change in Altitude

CO_2 has no such temperature limitations. At higher altitudes where water vapor is close to non-existent, CO_2 exists in similar percentages as near the ground, and this is where it dominates.

WARMING OF THE EARTH IS CAUSED SIMPLY BY THE HEAT GENERATED WHEN WE BURN THINGS.

It was hard to find any proof of this, for or against, so I did my own calculation. It was not hard. I found the amount of fossil fuels that have been burned over the last 75 years, and the associated heat generated, assuming a certain level of combustion efficiency. Then, knowing the mass and heat capacity of all the air on Earth, the following table was generated.

Identifier	Item		Units	Mathematical Operation
A	Time Span	75	years	
B	average consumption per year	7.00E+13	watt-hrs	
C	total consumption	5.25E+15	watt-hrs	A x B
D	total consumption	1.89E+19	Joules	Units Conversion
E	Mass of Atmosphere	5.15E+18	kg	
F	heat capacity per kg of air	700	J/kg°C	
G	heat capacity of whole atmosphere	3.60E+21	J/°C	E x F
H	Temperature Change	5.24E-03	°C	D / G

The estimated temperature change is only .0052°C, which can be considered negligible. This is not the mechanism that has been causing global warming.

WARMING IS CAUSED BY NATURAL FLUCTUATIONS IN ENERGY OUTPUT FROM THE SUN

It turns out that solar irradiance has been measured on a continuous basis. Irradiance is measured in units of power per unit of area, and it is the pure incoming radiation onto the earth without being polluted by the highly variable impact that the planet has on reflecting or absorbing it. Since 1978 measurements are based on observations from satellites in space. Prior to that, earth-bound instruments were used. There is also a historical record of sunspot activity going back to the 17th century, in other words since telescopes were available. Below is the data since 1978, considered to be the most accurate. I include the notes that accompany the graph, which explain the measurement process. (Kopp, 2016)

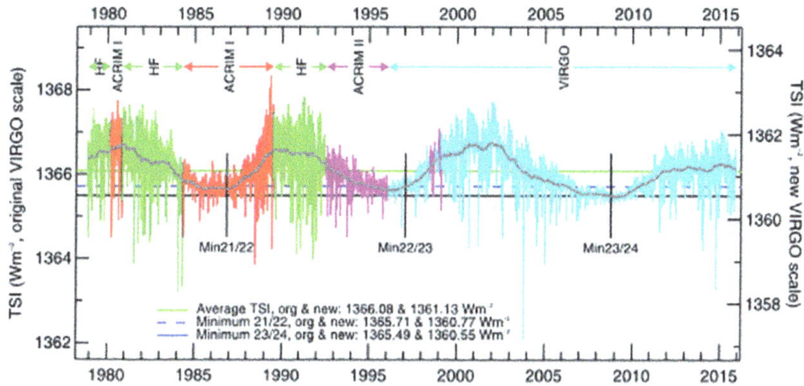

Figure 2: The PMOD TSI composite shows peak-to-peak TSI variability of ~0.1 % in each of the three solar cycles observed during the space-borne measurement record, with that variability being in phase with solar activity. The colors indicate the binary selections of different instruments used in the creation of the composite. The right-hand vertical scale indicates the more accurate currently-accepted absolute value. ("HF," short for its creators Hickey and Friedan, is a name used by some authors for the NIMBUS7/ERB instrument in Figure 1.) (Figure is courtesy of the VIRGO team)

Below is shown the long-term trend, where sunspot records are used with calibrated modern mathematical models to predict irradiance.

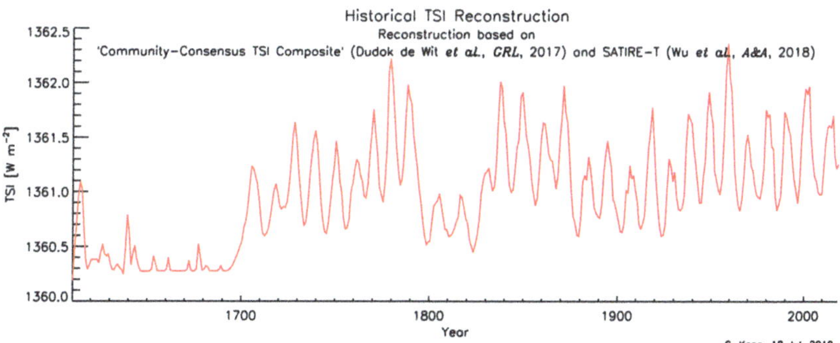

In both graphs, we can see a well-known 11-year cycle in the sun's output. The trend, however, has not changed in recent times, especially in the critical period since 1980 when global warming started to appear.

In 1997 a paper was published that used the theory of increasing solar output to explain the warming trend. The paper, called "Environmental Effects of Increased Atmospheric Carbon Dioxide", by Arthur B.

Robinson et al, showed the following graph. I include the caption below it to add context.

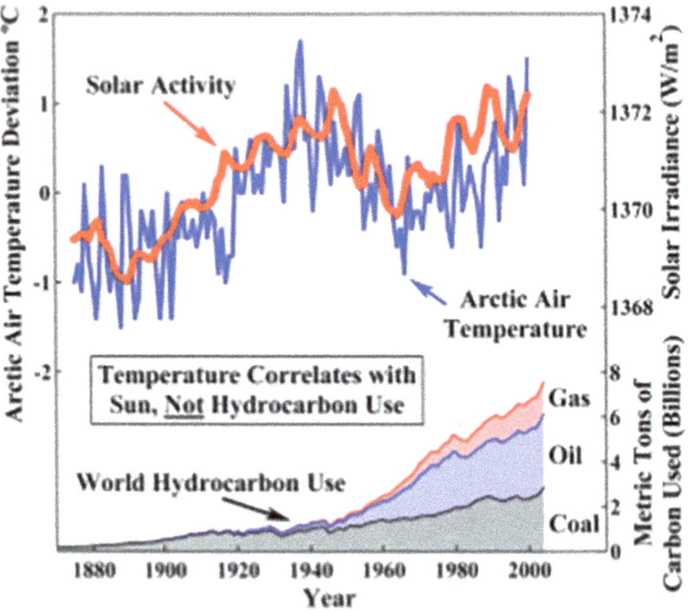

Figure 3: Arctic surface air temperature compared with total solar irradiance as measured by sunspot cycle amplitude, sunspot cycle length, solar equatorial rotation rate, fraction of penumbral spots, and decay rate of the 11-year sunspot cycle (8,9). Solar irradiance correlates well with Arctic temperature, while hydrocarbon use (7) does not correlate.

This graph shows a continuous increase in solar radiance going back to 1880, and especially a steep rise from around 1965 to 2000. The data does not agree with other sources, including those I showed previously. It also superimposes arctic temperature increase, rather than world temperature increase, presumably because there is a better correlation. The authors claim correlation of rising temperature with solar output, and none with CO_2 emissions. However, the basic data is faulty.

The same paper by Robinson, which was shown above to have faulty irradiance data, shows another graph claiming that ocean temperatures are in a normal increasing trend, which is cyclic going back centuries, see below.

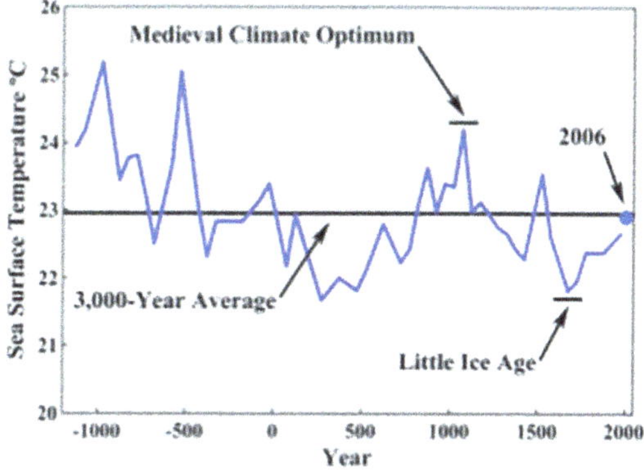

Figure 1: Surface temperatures in the Sargasso Sea, a 2 million square mile region of the Atlantic Ocean, with time resolution of 50 to 100 years and ending in 1975, as determined by isotope ratios of marine organism remains in sediment at the bottom of the sea (3). The horizontal line is the average temperature for this 3,000-year period. The Little Ice Age and Medieval Climate Optimum were naturally occurring, extended intervals of climate departures from the mean. A value of 0.25 °C, which is the change in Sargasso Sea temperature between 1975 and 2006, has been added to the 1975 data in order to provide a 2006 temperature value.

The raw data for this was generated by another author, Kiegwin, reported in a 1996 paper. It shows inferred temperatures in the Sargasso Sea, part of the Atlantic Ocean. The methodology used is rather cryptic, but basically residue of dead marine organisms found at the bottom of the ocean contain various isotopes of oxygen that can reveal temperatures experienced when the animals were alive. This graph is important because it began appearing in the News Media, and given its fundamental complexity, was accepted as fact. It was repeated and rehashed in various publications with intention of refuting that CO_2 emissions were the source of global warming. It was even used to generate the following satirical cartoon by cartoonist John Trever.

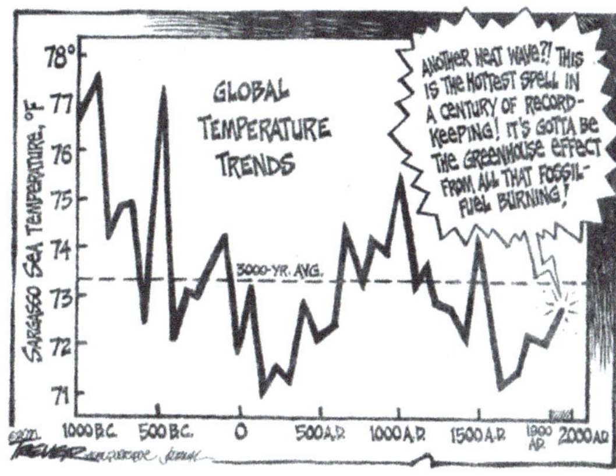

The Kieger paper was not intended to be an input to climate change debate, and if you look at the original one taken from that paper, you may get a different conclusion.

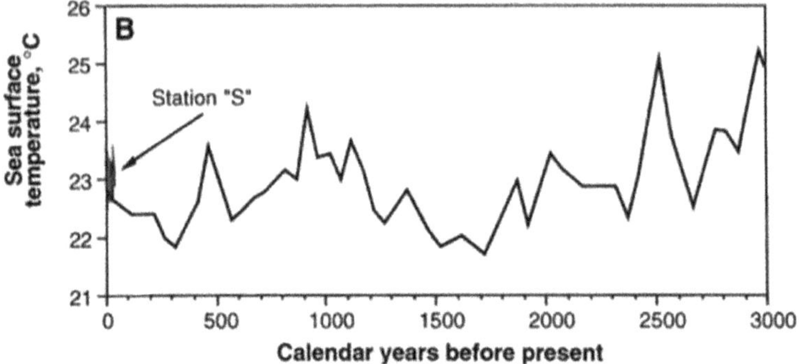

Firstly, the scale is reversed, so the beginning of the chart is present day. More importantly, there is a spike right at the beginning, identified as "Station S", and this represents temperature readings taken in the same sea since 1954, but with much more reliable modern instrumentation. It shows a big spike in temperature in the modern era, corresponding to the vast expansion of fossil fuel consumption. Somehow, this part of the data was left out of the Robinson paper.

CHAPTER CONCLUSIONS

In my search for "the truth", I have found that the data that claims warming of the planet as a result of greenhouse gases is credible. The raw data is mostly buried in very sophisticated published papers that are clearly the work of experts that are immersed into their particular sub-fields of research. They move along like all researchers, building upon previous work and gradually developing, as a community, an improved understanding of the subject.

On the other hand, I have found that those opposed to the greenhouse gas theory are most often lacking in credibility. There are many websites and chat rooms that consistently post inflammatory statements that are accompanied by either dubious science, or no science at all. I have not referenced them here, but I have attempted above to evaluate some of the skeptics' theories that have circulated and find them to be without basis.

This is not to say that some doubt should not remain. In my own experience, I have seen how brilliant and reputable researchers have managed to fool themselves, even inadvertently, in trying to match observations with the theory or the mathematical model that they are trying to develop. There is a strong driving force to get to the nirvana of science: a model that explains the past, and then, presumably, can predict the future. But some skepticism is always healthy and can lead to new insights.

For now, I am willing to declare the following conclusion:

There is a distinct warming of our planet that is not readily explainable except by increased emissions of greenhouse gases resulting from human activity. The level of increase in these emissions is at double the rate of population growth since the 1960's.

CHAPTER 2

THE CONSEQUENCES OF WARMING IN TODAY'S WORLD

THERE IS NO DOUBT THAT the terms "Global Warming" and "Climate Change" have made their way into the mainstream lexicon. It is generally accepted that Warming has arrived, and that consequences will probably be dire.

I stated in the introduction that I write this book as a moving journey of discovery, rather than as a record of foregone conclusions. I have to admit, however, that each time I hear a reference to Climate Change by a news broadcaster, a political pundit, or a politician, a wave of skepticism comes over me. This is not because I doubt the science, which is paramount. It is because it seems evident to me that the people using the terms, and their casual statements of certainty, are only repeating what they have heard, not what they have verified.

In this chapter, we will proceed as in the previous one. Having concluded that the world is indeed warming due to human-sourced (anthropogenic) greenhouse gases, we will next look at what the science says we should expect from Global Warming. Then, we will look at what evidence exists to confirm or deny what the basic science says.

There are 3 main areas to investigate: Extreme Weather, Habitat Loss and Sea Level Rise

1) Extreme Weather

Warmer Oceans will more readily allow water to evaporate into the atmosphere, and warmer air will more readily absorb it. Although relative humidity may stay constant, the specific humidity will increase. This means that the mass of water in the air will be greater. When warm air rises, water vapor encounters the cooler temperatures at the higher altitudes and begins to condense, forming clouds. This leads to more rain.

The condensation of water also releases latent heat into the upper atmosphere. In the same way that it requires heat input to evaporate water, heat is released when it condenses back to water.

The result is more water and more heat in the upper atmosphere. Storms are created when a mass of colder air approaches, and the temperature difference between cold and warm air causes instability. In the extreme cases, the collision of cold and warm air causes the swirling motion that we see in hurricanes and tornados, courtesy of the Coriolis effect.

It could be argued that the approaching colder air in the above scenario is also warmer than it would have been prior to global warming. The temperature *difference* between cold and warm air might not be changed, and it is the difference that creates the instability. The wild card, however, is latent heat. The temperature difference should still be greater because the heat released from the condensation process is superimposed onto the warm side of the equation.

In regions of the earth that typically see extreme storms, it can then be expected that they might become more extreme.

Increase in inland rainfall can cause overflow of riverbanks and local flooding, while tropical storms can cause temporary flooding in coastal regions.

The science presented here is admittedly simplistic when compared to all of the factors that influence weather. We will see later when we look at real world data if it applies.

In dry regions of the earth, such as desert climates, higher air temperatures should result in increased evaporation of what little water there is in the soil and plant life. Arid climates would then be expected to get even dryer, which would negatively affect wildlife and agriculture.

2) Risks to Wildlife Caused by Habitat Change

It is difficult to describe the science here because each species has its own vulnerabilities. Those sensitive to ambient temperature would be clearly at risk. Secondary effects, such as melting of glacial ice and sea ice, could also affect certain marine animals, and land animals dependent on ice formations. Incursions of wildlife into warmer climates where they previously did not exist can also wreak havoc on food chains.

It may be best to address this category through examples and observations, which we will do further along in this chapter.

3) Rising Sea Levels Caused by Melting of Glacial Ice and Water Thermal Expansion

Near the poles, and on mountains at high altitudes, there are massive stores of water in glaciers. Glaciers are ice sheets that survive summer warming because of their depth and breadth. They resist seasonal melting in all but their outer exposed layers. They also get replenished in winter months when snowfall gets compacted into new ice.

The size of any glacier can be said to stay roughly constant as long as average temperatures also stay constant over long periods of time. In the Earth's history, there have been slow warming and cooling periods, lasting hundreds of years or even millennia. In these periods, glaciers have shrunk or grown according to the climate.

In our time the average temperature increases of around $1°C$ that we described in the previous chapter has taken place over a truly short period of time as compared to ancient historical trends. We might therefore expect that glaciers may be affected in a way that is visible to us right now.

Warming of the oceans will also increase seal levels all by itself, even without considering ice melt. All objects expand when heat is added and contract when heat is lost. The upper regions of the oceans that get warmer will naturally take up more space, causing sea levels to rise.

Now, let us see what relevant observations we can find in today's world.

EXTREME WEATHER

In 2014 a US Federal Advisory Committee issued "The National Climate Assessment" (Melillo, 2014), a comprehensive study intended specifically to report on the effects of Global Warming in the United States. There is some global data in there, but mostly it reports observations in USA and its coastlines.

Below I include a lengthy extract from the overall summary given in the report:

"Americans are noticing changes all around them. Summers are longer and hotter, and extended periods of unusual heat last longer than any living American has ever experienced. Winters are generally shorter and warmer. Rain comes in heavier downpours. People are seeing changes in the length and severity of seasonal allergies, the plant varieties that thrive in their gardens, and the kinds of birds they see in any particular month in their neighborhoods.

Other changes are even more dramatic. Residents of some coastal cities see their streets flood more regularly during storms and high tides. Inland cities near large rivers also experience more flooding, especially in the Midwest and Northeast. Insurance rates are rising in some vulnerable locations, and insurance is no longer available in others. Hotter and drier weather and earlier snow melt mean that wildfires in the West start earlier in the spring, last later into the fall, and burn more acreage. In Arctic Alaska, the summer sea ice that once protected the coasts has receded, and autumn storms now cause more erosion, threatening many communities with relocation.

Scientists who study climate change confirm that these observations are consistent with significant changes in Earth's climatic trends. Long-term, independent records from weather stations, satellites, ocean buoys, tide gauges, and many other data sources all confirm that our nation, like the rest of the world, is warming. Precipitation patterns are changing, sea level is rising, the oceans are becoming more acidic, and the frequency and intensity of some extreme weather events are increasing. Many lines of independent evidence demonstrate that the rapid warming of the past half-century is due primarily to human activities."

This seems to be damning stuff. Let us examine some of data reported.

MORE POWERFUL STORMS

The chart below shows North Atlantic and Pacific storm severity covering the years 1970 to 2009. I have left the caption to provide better understanding.

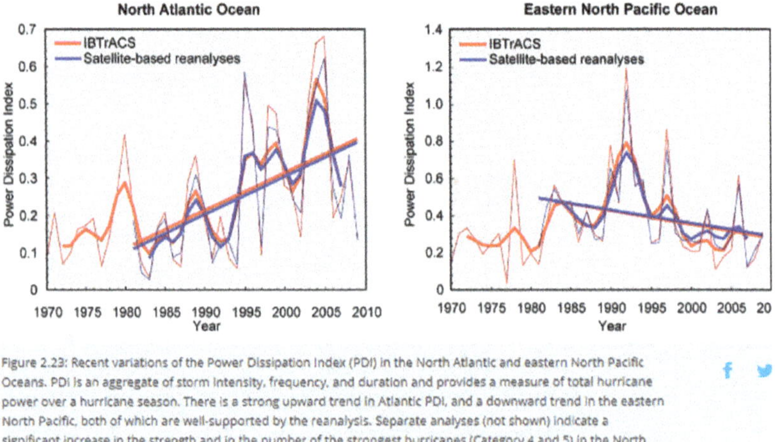

Figure 2.23: Recent variations of the Power Dissipation Index (PDI) in the North Atlantic and eastern North Pacific Oceans. PDI is an aggregate of storm intensity, frequency, and duration and provides a measure of total hurricane power over a hurricane season. There is a strong upward trend in Atlantic PDI, and a downward trend in the eastern North Pacific, both of which are well-supported by the reanalysis. Separate analyses (not shown) indicate a significant increase in the strength and in the number of the strongest hurricanes (Category 4 and 5) in the North Atlantic over this same time period. The PDI is calculated from historical data (IBTrACS[11]) and from reanalyses using satellite data (UW/NCDC & ADT-HURSAT[1,12]). IBTrACS is the International Best Track Archive for Climate Stewardship. UW/NCDC is the University of Wisconsin/NOAA National Climatic Data Center satellite-derived hurricane intensity dataset, and ADT-HURSAT is the Advanced Dvorak Technique–Hurricane Satellite dataset (Figure

There are a few things to consider before concluding on the meaning of these graphs. First, the PDI index is a measure of both maximum wind speeds and the number of times they occur over the hurricane's life. I noticed, however, that the index uses speed to the third power (V^3). This is curious, because both the force exerted onto an object by the wind, and the energy contained in the air would be a function of speed squared (V^2). So, the index will tend to exaggerate any upward or downward trends. Additionally, the number of times a certain wind speed occurs is not a measure of total energy over time.

Secondly, the time period is restricted even though data exists over longer periods. If we are looking for trends associated with things like CO_2 emissions and industrialization, we need to look further back.

Thirdly, looking at the regions encompassed by the United States does not necessarily tell us what we need to know about global climate.

The following chart found on the website of the Cato Institute shows data going much further back in time, and also forward to 2017.

North Atlantic Hurricane Intensity (PDI), 1920-2016

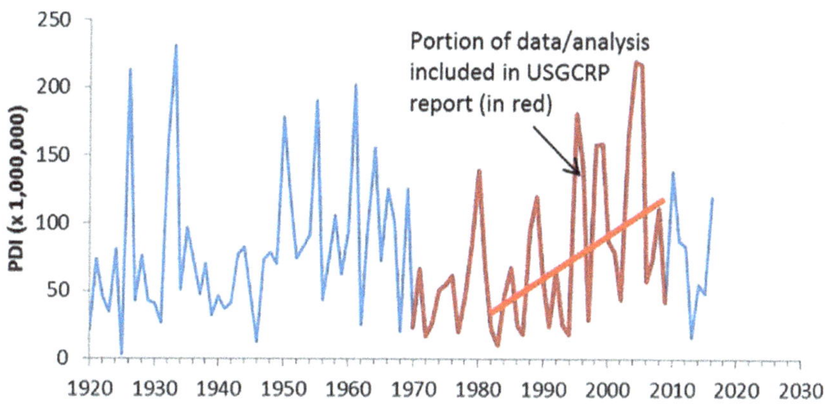

There is pronounced skepticism and irony included in the graph's description, suggesting the data from the Climate Assessment was cherry-picked. Being skeptical myself, I checked other sources (National Oceanic and Atmospheric Administration) and found that the chart above is valid. Over this longer longer time, and considering more recent storms, it certainly cannot be concluded that severity of storms has increased. It should be noted that data pre-1970 did not have satellite observations included and therefore could be considered less accurate.

Meteorologist Dr. Ryan Maue has published more complete data covering the whole globe and using a different index that uses wind speed squared instead of cubed (Accumulated Cyclone Energy, or

ACE Index). Below is a chart containing data in the Satellite era (from 1970).

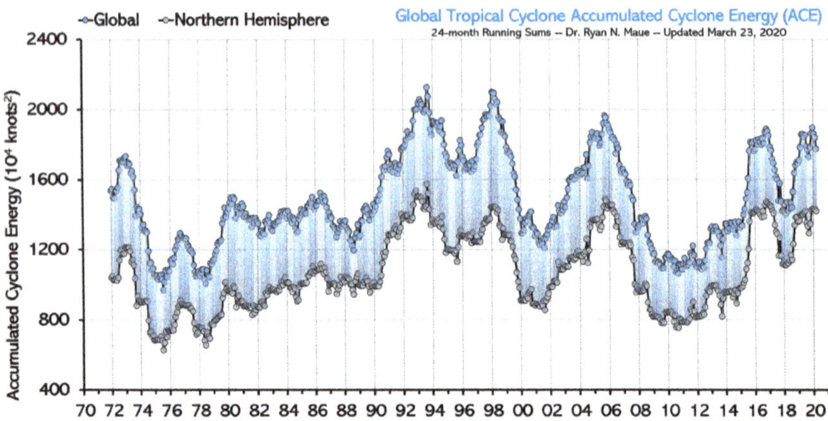

Globally there is no obvious trend that we could relate in time to the warming of oceans and air that we saw in the previous chapter. In terms of Hurricane frequency, the chart below shows, if anything, a decreasing trend in number of hurricanes with wind speeds greater than 64 knots, and a stable trend for those with higher wind speeds.

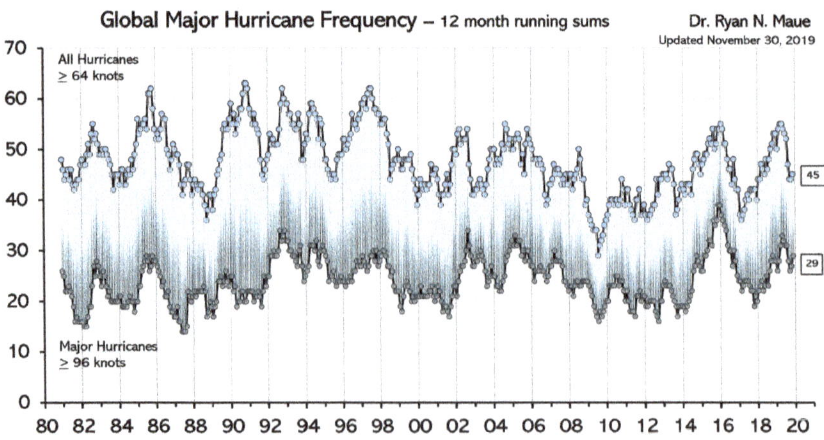

As another confirmation, the NOAA states the following in a highly informative study on hurricanes: "In short, the historical Atlantic hurricane frequency record does not provide compelling evidence for a substantial greenhouse warming-induced long-term increase." They use the following interesting graph as evidence.

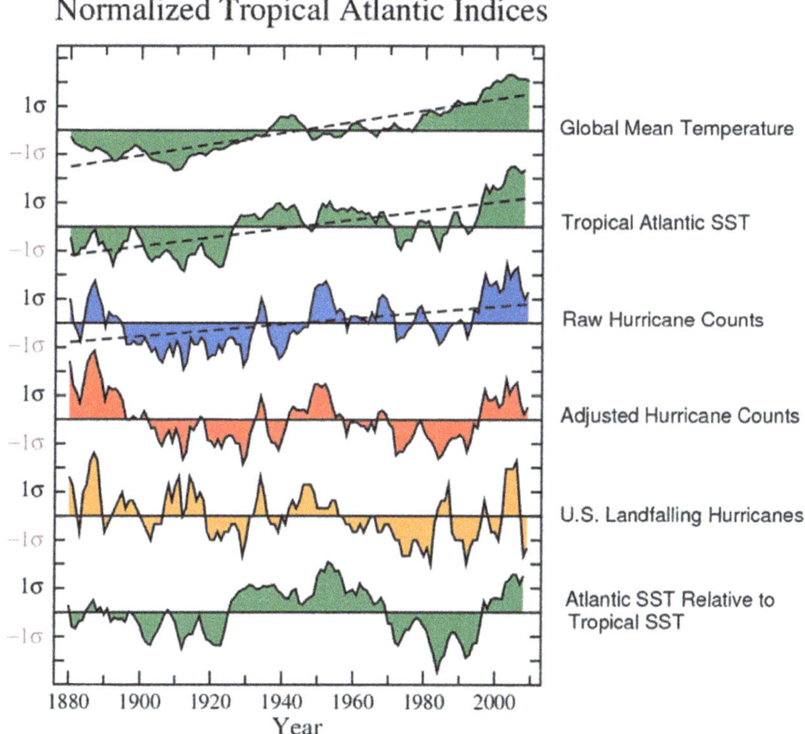

In this graph, the colors indicate deviation from an average taken over the whole scale shown. Both air and ocean temperatures show the rising trend previously discussed, but hurricane counts show no similar trend (red and yellow graphs).

This is not to say that Global Warming cannot have an effect on storm severity. Rather, it says that, *so far*, other more powerful factors are still dominating the stage.

INCREASED PRECIPITATION

It turns out that determining levels of precipitation on a global scale is exceedingly difficult. Obviously, rain does not fall uniformly over the surface of the planet, or even uniformly inside one city. Experts take massive amounts of individual rainfall data points and then perform some math to come up with globally averaged amounts, and it is certainly inexact. Some scientists have taken to indirect measures like ocean salinity, which we will describe later in the chapter. The key is to look for trend, presuming that the trends are right even if the absolute value of rainfall is "inexact".

The 2014 Climate Assessment report also weighs in on rainfall in Continental USA. See Chart Below.

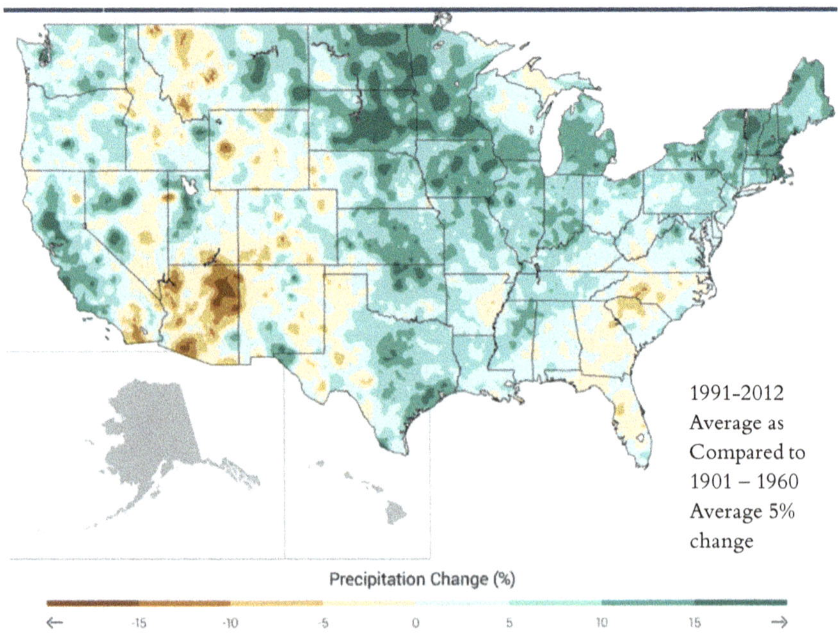

There seems to be an increase of rainfall, on average 5%, since 1991. The chart also shows that the increases are in normally "wet" areas, and the decreases are in normally "dry" areas. This is consistent with the basic science which says that wet areas get wetter and dry areas get dryer.

The following global data on rainfall covers the era 1900 to 2005. (Li Ren, 2013)

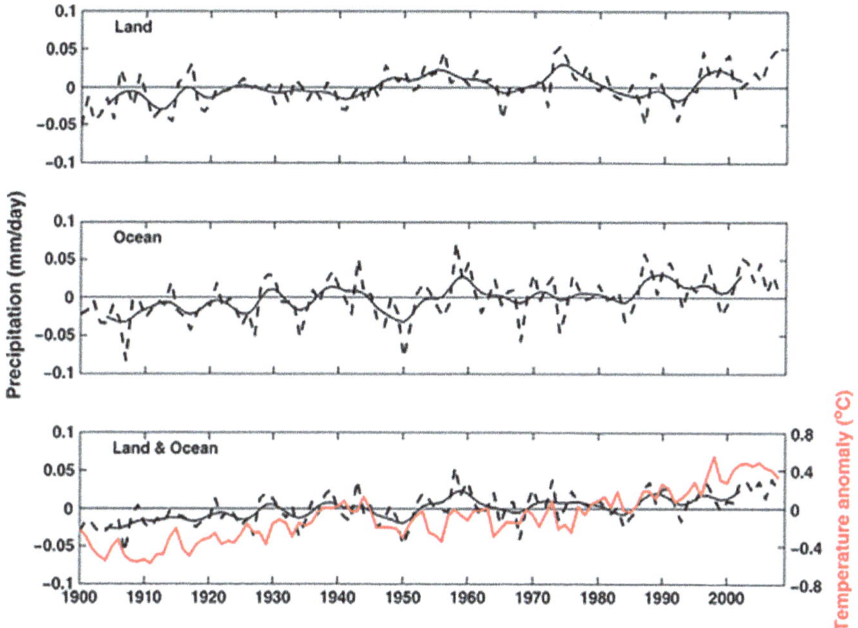

These graphs are plotted as rainfall anomaly as compared to the average of all the data going back more than 100 years. In other words, if in any given year the amount of rainfall is exactly equal to the average historical amount, then the anomaly will be zero.

In the lowest of the 3 graphs, total rainfall trend can be seen on the same timeline as SST (Sea Surface Temperature) increase, consistent with Global Warming. There is a trend of increased rainfall from 1900 to 1940, then a stable period up to 1980, and then perhaps a small rising trend after that. As it happens 1980 was a pivot year where global temperatures started to have a noticeable rise.

In this paper the authors conclude that a there has been a trend of rainfall increase of .04mm/day per 100 years over the oceans, and .03 mm/day per 100 years over land. This means that it has taken 100 years to see an increase of .04 mm/day. This conclusion comes from some sophisticated smoothing analysis of the bumpy raw data. I question its validity for our purposes. We are not much interested, for example, in the rising rainfall trend in the early part of the 20th century because that era did not have any greenhouse gas-induced warming. Yet it colors the conclusion in a profound way because it is visually evident that increases since then have been less pronounced. It is interesting, though, that rainfall and temperature followed each other quite well in those early years, even if the warming was caused by other factors.

It is important to know what this increase represents as a percentage of "normal". I found raw rainfall data in a different study where the authors use a data set from the Radcliffe Observatory at Oxford. (Fubao Sun, 2018). This graph is raw data of global land-based rainfall. It certainly does not appear to show any upward trend. In fact, some very sophisticated statistical work done in the paper concludes that precipitation in the 2000 to 2009 decade was indistinguishable from random as compared to the behavior since 1940. Year-to-Year variation dominates here, and there is no upward trend. In order to correlate with the chart from the Ren paper above, notice that there is a distinct upward trend in the 1900 to 1940 time period in both datasets.

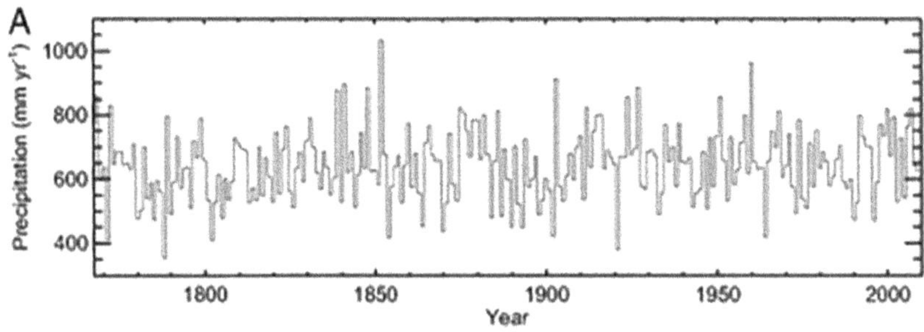

From this graph we can extract roughly the average, or "normal", amount of rainfall. It is approximately 650 mm/yr. If we take from the Ren paper the largest, and probably exaggerated, rising trend of .03 mm/day per 100 years over land, then we could expect an increase of .012 mm/day over the 40-year period from 1980 to 2020. That equals 4.38 mm/yr, or 0.7% increase in 2020 as compared to 1980. This number is so low it would certainly be lost in the natural year-to-year variation.

The graphic below is taken from the Ren paper. It shows rainfall anomaly from 3 different datasets, with the one identified as "REC" being the most recent reconstruction of historical data. It shows that most of the anomaly occurs over the oceans. Since the increased rainwater originates from the ocean, and then mostly falls back into the ocean, it is not expected that there is a significant trend on land, where flooding could have been affected.

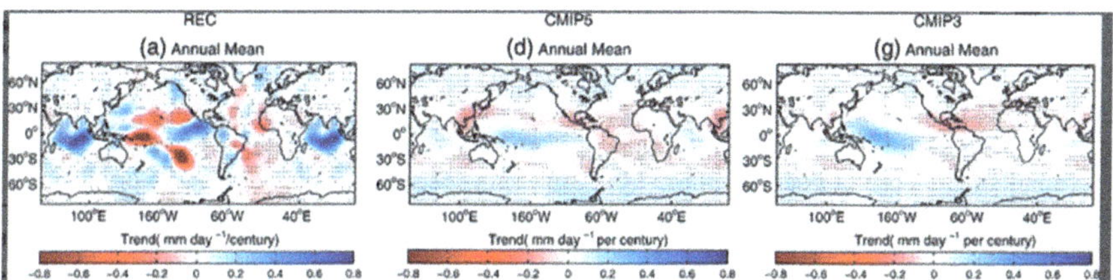

Returning to the 2014 US Climate Assessment Report, how do we reconcile the above underwhelming trends data with the claim that the continental US has experienced an average of 5% increase in rainfall between 1990 and 2012? I went to find the raw data for US rainfall history. (Wang, 2020)

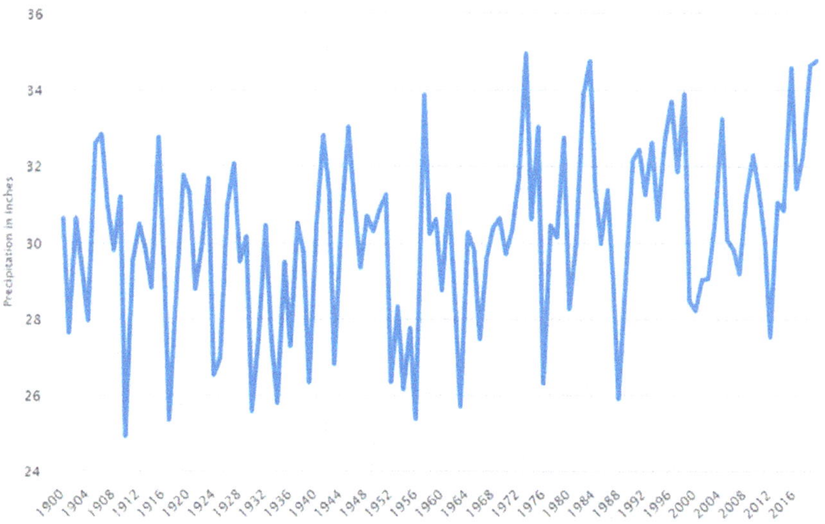

Recall that world temperature trend was stable before 1980, and then started a 40 year upward trend. So, I took the average of the data from 1980 to 2019 and compared it to the average of the previous 40 years, 1940 to 1979. I find an average increase of 3.8%. It is also somewhat visible in the graph that there has been a shift. The Climate

Assessment report concluded 5% because they used a 100-year average as a baseline, which included that early 20th century era of low rainfall not related to greenhouse gas warming.

In the end, the global numbers should dominate over any local phenomenon, but we can probably still conclude that warming has increased rainfall in USA.

Looking at still other sources, we find some contradictory data. (Huihui Feng, 2015) Satellite imagery of soil moisture content shows the DGDWGW paradigm (dry gets dryer wet gets wetter) is not supported. Only 15.1% of land surface has followed it, while 7.8% have followed the opposite trend.

This is roughly in agreement with a 2014 paper (Peter Greve, 2014) which showed 10.8% following the trend and 9.5 exhibiting an opposite trend, using hydrological data from 1948 to 2005. Note that these two studies use entirely different types of data, that is, physical land measurements, as compared to the satellite imagery data, and both support the same conclusion.

In another paper the authors conclude the opposite. (Nikolaos Skliris, 2016) Using ocean salinity data, they conclude there is increased rainfall. Ocean areas near outflow of rivers are less saline, indicating more fresh water entering than before. Mid-Ocean areas are more saline, indicating more evaporation into the atmosphere, and thus more rain. This is of course an indirect measurement but still an interesting finding.

There is a large variety of data sources and they are certainly somewhat contradictory. I have to conclude that, so far, if there is an increase in precipitation it is marginal and perhaps not especially important in terms of effect on humans.

EFFECTS IN ARID ENVIRONMENTS

Some global data discussed above showed that the trend towards dry areas being drier as a result of global warming was weak, and in some areas even contra-indicated. Here I choose to examine two specific dry environments to see what has been happening with respect to global warming. The Australian Bush Fires and the California wildfires have garnered a great deal of attention in the media. Let us have a look.

1) Australian Bush Fires

Tennis fans who were watching the Australian Open in January 2020 got a daily dose of updates on the bush fires that were causing so much smog that some of the matches had to be cancelled. Athletes were helping donating money towards relief efforts. In the news, it was certainly a foregone conclusion that the fires were a direct result of global warming.

In this section we will do as we did in previous ones, look at historical data regarding these fires to see what the trends show. Firstly, Australia has seen its share of temperature increase, in line with the global average of around 1°C. The chart below represents data from ACORN-SAT, Australia's own land-based measurements from 112 locations around the continent.

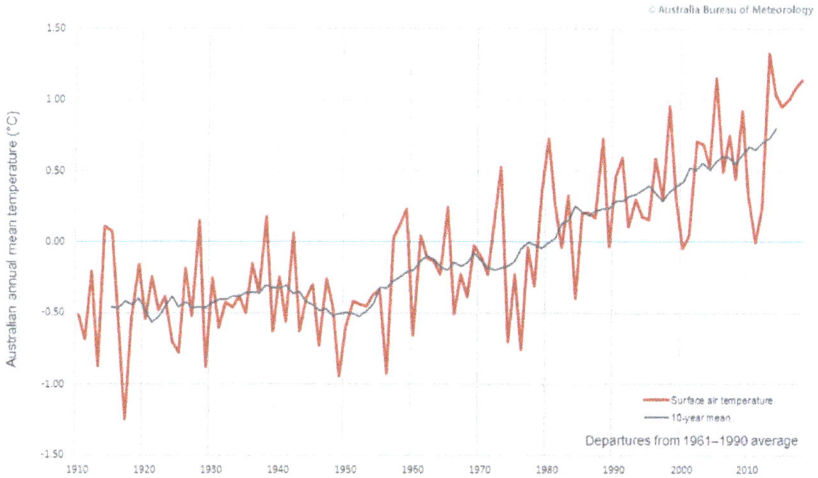

Australian weather is notoriously volatile. There are big swings from year to year in rainfall, for example. The continent does not have many tempering features to smooth out temperature and precipitation variance. There is a complete lack of inland lakes and rivers, and no mountainous regions that would otherwise produce snow melt.

It could be expected that a change of 1°C would be sufficient for dry areas to lose even more moisture. In regions categorized as arid, rainfall is insufficient to replenish moisture lost to evaporation. Surprisingly (at least to me), the bush fires often occur in temperate regions near the coast lines. The interior regions, which are classified as desert, also experience fires but there is less vegetation to act as fuel.

Below is an image of the bush fire status in Victoria and New South Wales in early January 2020.

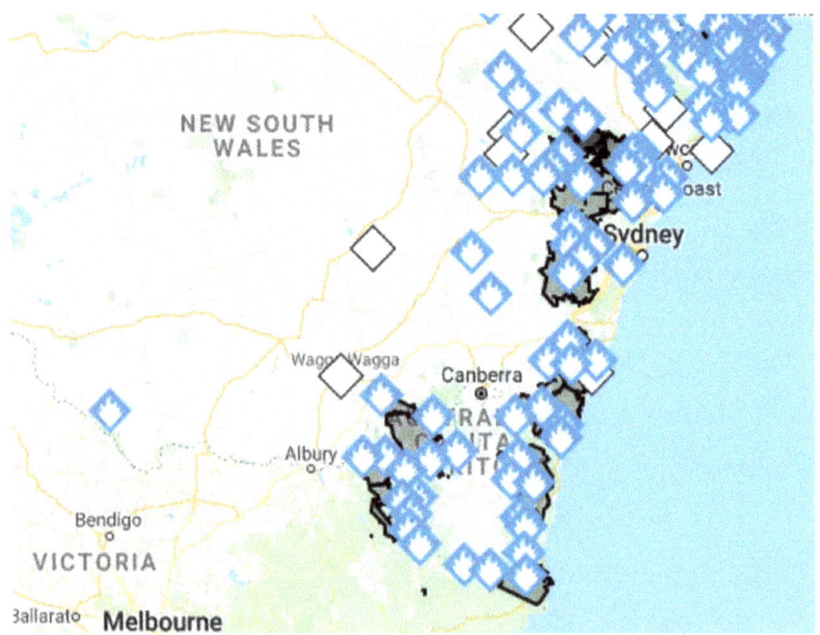

Image Issued by New South Wales Government Jan 3, 2020

Since the fires were raging near Canberra and Sydney, I decided to look at historical data in those areas. The Australian Bureau of Meteorology conveniently provides detailed precipitation levels at individual measuring stations. I was able to find data for Stations 070000 and 066062, close to Canberra and Sydney, respectively.

There is no obvious rainfall trend in either increasing or decreasing directions. The temperature escalation started in 1980, but over the next 40 years we don't see any trends in rainfall. In fact going back around 100 years, you can find years of low and high precitptation, indicating that volatility has not changed in that time period.

It's actually difficult to know what to expect since these are not categorised as arid areas. It is in line, however, with previous observations in this chapter which lead to a conclusion that average rainfall has so far not been significantly affected by global warming.

The last points in time on these charts show data for 2019. It can be seen that this was an exceptionally dry year.

Region	Canberra	Sydney
Historical Average – Annual	638	1205
2019 Average – Annual	409	852
Percentile	8.5	11
Historical Average – September to December	236	315
2019 Average – September to December	83	175
Percentile	1.3	16

Percentiles are taken from statistical analysis of the data. Taking Canberra rainfall in the spring and summer months from September to December as an example, 1.3 percentile means that over a 100-year period, you could expect only 1.3 years to be this dry or dryer. This naturally goes a long way to explaining the fires. The vegetation would be very dry and would readily succumb to initiation and propagation of fires, which are almost exclusively started by lightning.

It is still impossible, however, to see a trend over long time scales, which would be required if we are to assert that these 2019-2020 fires are a result of global warming. Looking at the graphs of raw data, it can be seen that even in the 3 years prior to 2019, there was ample rainfall, in line with historical averages.

But is rainfall the key? I looked at detailed data for the state of Victoria, going back to late 19th century. I chose 2 stations, one in Melbourne near the ocean, and one in Bendigo which is inland and much drier. I found 15 important years for bushfires. When I average the rainfall for all of those years in the 2 months preceding the fires, it is almost exactly half of the historical average over all years. I guess it is obvious that less rain means drier plant life and more chance of fire. The important finding, however, is that similar to the graphs shown for Canberra and Sydney above, there is no historical trend leading to less rain in the recent years of global warming.

In fact, the theory predicts more rain, as we discussed at the beginning of this chapter. Warmer oceans and the warmer surrounding air are supposed to lead to more rain. But the Australian data shows no such trend, which agrees with most of the rainfall studies already discussed.

What about the warmer air's increased ability to absorb water, thus causing drier vegetation? The theory says that as the air temperature gets warmer, relative humidity would stay the same, while the mass of water in the air would increase. However, let us look at the worst-case scenario from a fire point of view: suppose that the water content stays the same, while the temperature goes up. In this case I choose a reference relative humidity of 30% at 30C. If the temperature rises by 1°C, holding water content constant, then relative humidity goes down to 28.4%. This results in the air having a 2% greater ability to absorb water from the vegetation. Not too dramatic, but this could have an influence.

The 2019-2020 fires gained a great deal of attention mainly because of the locations, in temperate areas close to the ocean. This resulted in powerful effects on humans, and loss of life and habitat for millions of animals. In this sense, it was truly unprecedented. It was more important than the largest bushfire in history in 1974, which burned mostly inland desert areas almost devoid of life. From the point of view of attributing cause, however, we can only say that 2019 was an exceptionally dry year, but not unprecedented, and not yet part of any historical trend.

One thing is undeniable. Looking at it in the simplest of ways, fires are almost non-existent in winter where mean temperature on the continent is around 10°C, while fires occur every year in summer, with a mean of 27°C. The 17°C increase can therefore be concluded to be the root cause of all fires. A further increase of 1°C, or even more in future, can then be assumed to be an aggravating factor for Australian bush fires, even if trend data so far does not show it.

2) California Wildfires

Much of the State of California has a desert climate. If you have travelled in California, you will notice that you don't have to go too far inland to see the weather change from temperate to hot and dry.

California has a population of 39 Million, which is slightly greater than all of Canada. With coastline property increasingly at a premium, many communities have formed inland where the conditions are conducive to fire. There are 2.7 Million people living in zones categorized as "very high fire hazard severity zones".

California has a long history of fire management, and wildfires are nothing new. Fires are deliberately started outside of the fire season to reduce the quantity of "fuel", with the expectation that damages in the following fire season can be attenuated.

Fires are aggravated by strong winds. In California they have names: "Diablo" in the north and "Santa Ana" in the south. Fires also create their own winds due to the strong temperature gradients around them and the forced influx of oxygen that feed them.

The years 2017 and 2018 were difficult. The November 2018 "Camp" Fire (named after Camp Creek Road where it started) was considered the largest natural disaster in the world for that year.

While most of the Australian fires are started by Lightning, the large majority of California fires are started by human activity. This includes the downing of power lines, the cause of the Camp fire.

A comprehensive paper was published in early 2018 (Jon E Keeley, 2018) describing the sources of ignition, with a view towards better fire management in the future. Data is presented over a hundred-year span, and records were sufficiently detailed to identify extents of damage and sources of ignition.

In many of the graphs presented in this paper there is an apparent pivot point in 1980. Fires appear to be *less* frequent and less damaging after that year, which is coincidently the year when warming of the planet began. The reasons do not appear to be climate-related, however. More likely explanations are given as improved fire management practices and greater awareness of ignition sources among the population.

Using statistical analysis of the data, and bringing climate conditions into the analysis, there was a finding that prior year's rainfall correlated with larger fires. In this case, *larger* amount of rainfall was the culprit, contrary to the negative effects of *reduced* rainfall that we saw in Australia. When there is a large amount of rain, more vegetation is produced, and this becomes the fuel for the following year's fires. This illustrates the difficulty in trying to simplistically relate fires to climate change.

In the USA, the Palmer Drought Severity Index (PDSI) is used as a tool to compare regions and time periods. It takes readily available precipitation and temperature data and is produced with calculations that include evaporative effects. It does not, however, consider delayed sources of water, which in this case is mainly snow melt. The following charts were produced (Anna L. Jacobsen, 2018) show PDSI values for the South Coast and Sacramento Basin.

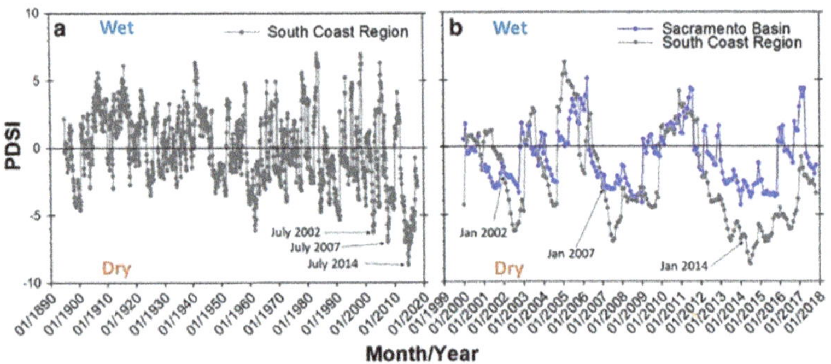

Zero on the graph represents the average of all the data, negative indicates drier than average, and positive indicates wetter than average.

Since 2000 there is a pretty clear trend towards drier climate, especially when looking over the full 120 years history on the South Coast. Can this trend be attributed to Global Warming? Possibly, because it corresponds well with the period of warming.

Below is a chart showing rainfall in Los Angeles with data going back to 1878. Each vertical bar represents one year, with 2018 on the far left and 1878 on the far right.

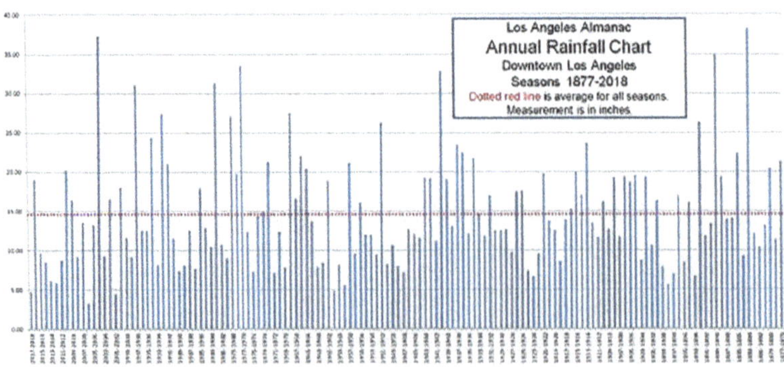

Starting in 2012 we see 4 years in a row of precipitation below the average (red line). We also see a span of 7 such years in the 1940's, and many other spans of 2 to 4 years. The PDSI data above is perhaps more dramatic because it considers the amount of sunshine causing evaporation.

In 2020 wildfires returned with even greater impact in terms of total area affected. As I write this fire has raged over 16000 square kilometers, which is a record and more than double the previous record set in 2018. Management of the crisis by authorities has fortunately led to fewer deaths and fewer structures affected. Still, the scale of this event must almost certainly lead to the conclusion that the trend towards drier conditions and larger fires has likely been greatly affected by human-induced Climate Change.

A snapshot of the drought status for the entire USA is shown below, for April 20, 2020. It is taken from a web site called "United States Drought Monitor" which keeps continuous track of the issue. Indeed, it shows Northern California to be in a state of severe drought. This is important for the whole state because most of the water reservoir capacity is more in the North and are fed not only by rainfall but by snow melt from the Sierra Nevada Mountains. Other areas in the South West are also experiencing drought.

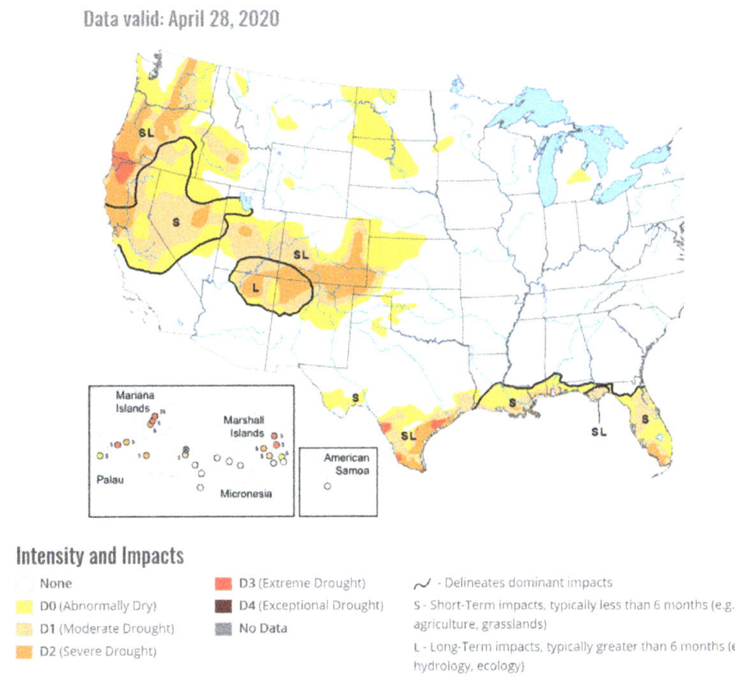

Recall that earlier in this chapter, we concluded that in the Northeast of USA there has been around 3.8% increase in rainfall over the last 40 years. Now we see South West areas being in frequent state of drought. Perhaps the USA is complying with the WGWDGD paradigm (Wet Gets Wetter Dry Gets Dryer), even though surveys on a global scale do not support this.

Here is one more piece of evidence that the US Southwest is getting hotter and drier.

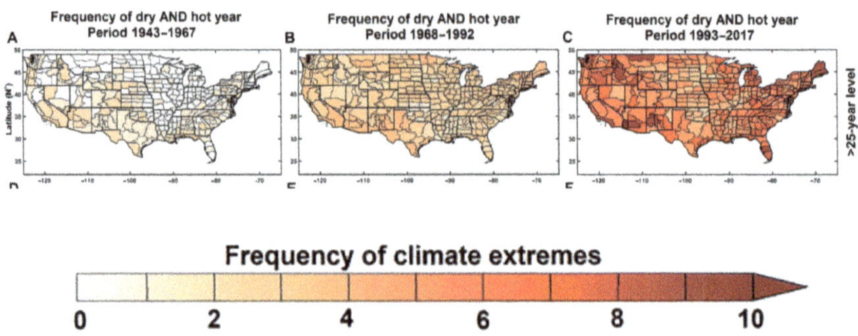

In this study (Mohammad Reza, 2020) The colors indicate the number of years that the weather was both hotter and drier than the statistical extreme of the previous 25 years. This means that if things are going along without any important changes, the color should reflect the number 1 on the scale provided. Over time, there has been a significant shift towards hotter and drier. This evidence is more scientific than looking at recent singular events because it shows a trend building up over a long period of time.

3) The Nile Basin

The Nile River, longest in the world, crosses through 10 countries, and approximately 270 Million people live there and rely on the river water for drinking, agriculture and hydro-electric power. Surrounding areas are mostly considered arid or semi-arid climates. It is a region that, even in the eras prior to global warming, has shown itself to be susceptible to both drought and flooding, depending on the level of water supply to the river.

The region has seen similar level of temperature change that we have seen elsewhere.

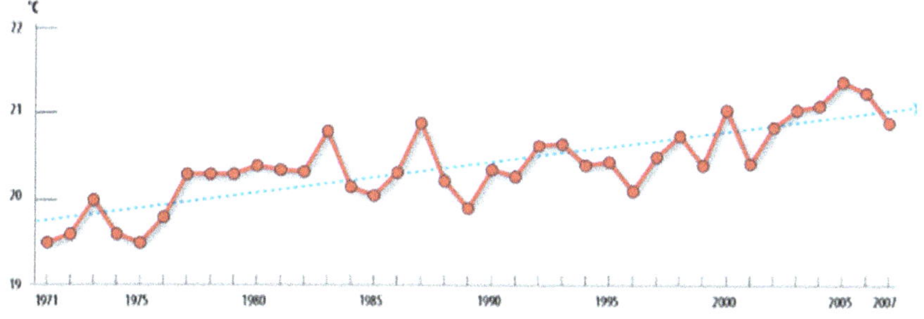

The Nile is mostly fed by equatorial lakes, the largest of which is Lake Victoria. In the image below, historical river flows are shown at various hydrological measurement stations. This is taken from The Nile Basin Water Resources Atlas which can be found online.

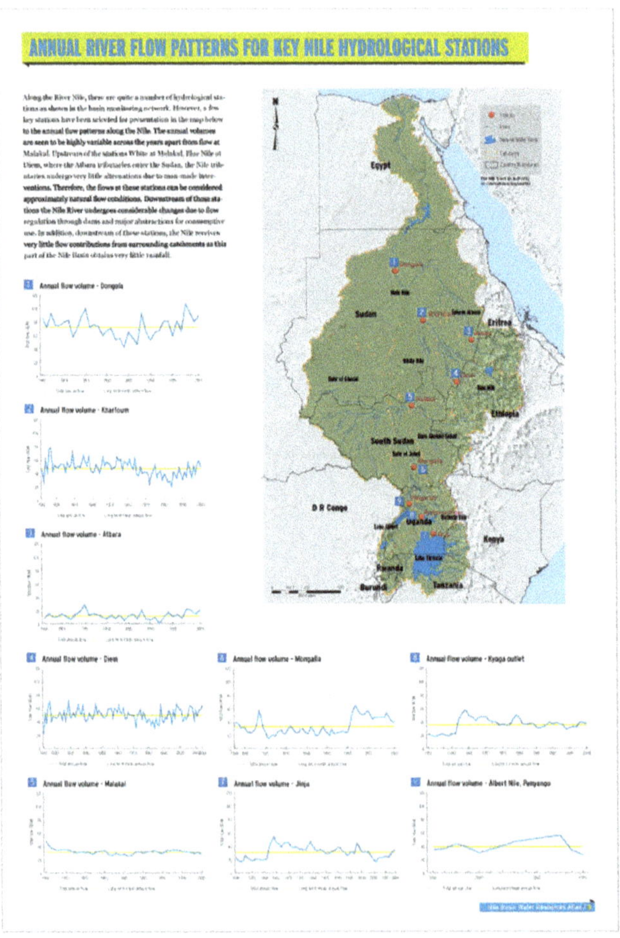

The stations in this figure were chosen to represent "natural" flows, that is, those that are not affected by human interventions, such as dams. It can be seen that historical flows have natural year-to-year variations, but no trend emerges in the global warming era, post 1980. There are also rainfall statistics available around the equatorial lakes that feed the river, but I do not present any details here because there are simply no trends to see so far.

I present this information specifically because, in this region where climate change would have dramatic effects on human life, there are so far no important observations to make.

The Nile Basin Initiative is a consortium of the 10 affected countries with the following vision: '*To achieve sustainable socio-economic development through the equitable utilization of, and benefit from, the common Nile Basin water resources.* On their website that describe many initiatives intended to pro-actively stave off the possible future effects of climate change. There is no evidence, however, of important effects in current times.

A CONTRARY OPINION ABOUT EXTREME WEATHER

In 2020 the United Nations (UN) issued a report called "Human Cost of Disasters" (Joris van Loenhout, 2020). Many media articles followed claiming that most of the costs of disasters, both human and financial, were the result of climate change. In the introduction the authors state that "this report focuses primarily on the staggering rise in climate-related disasters over the last twenty years". No wonder alarm bells were rung.

Let us look at some of the details. Here is a chart summarizing disasters of all types. Disasters that might be caused by Climate Change are in the categories Hydrological (floods), Meteorological (Storms) and Climatological (Fires and Droughts).

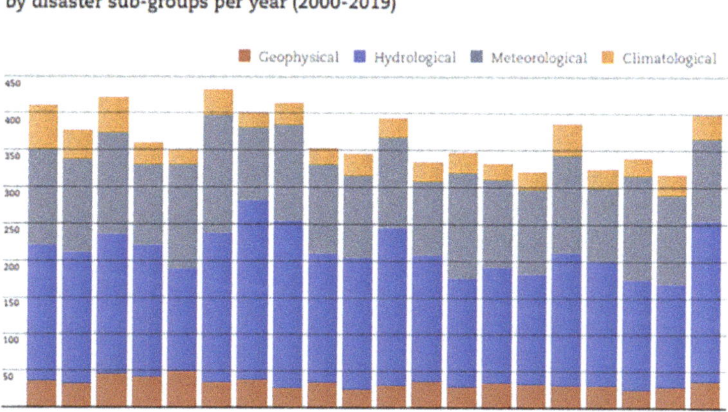

Although the authors state that the total number of events is much greater than in the previous 20 years, there is certainly no increasing trend in the 20 years in which this study was done. If anything, there is a decrease.

Drilling down a bit further, here is a chart showing country of occurrence, summed over the whole 20 years.

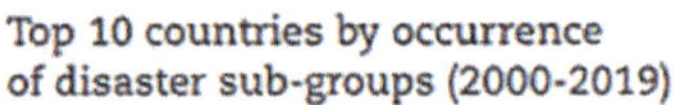

**Top 10 countries by occurrence
of disaster sub-groups (2000-2019)**

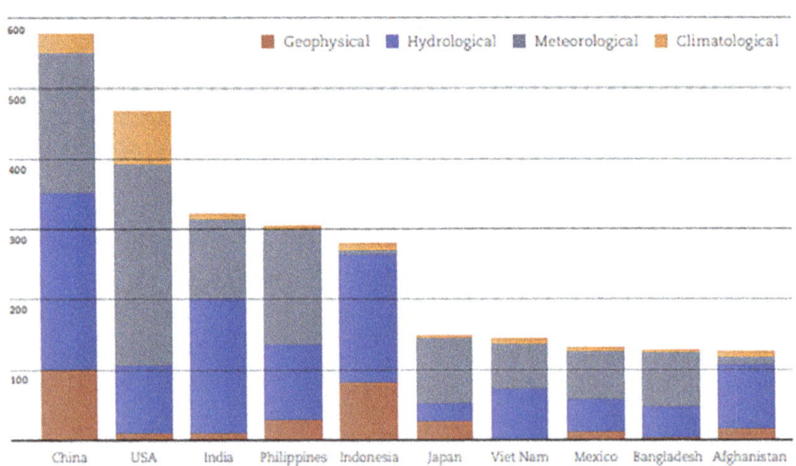

We see that the USA has the highest occurrence of storms, undoubtedly influenced by the Atlantic hurricanes and tropical storms already discussed. Floods occur mostly in Asian countries where population explosion has certainly put more and more people along critical waterways. Fires and droughts are mostly in USA, and this we have already discussed and shown that Climate Change is an important contributing factor.

Now let us look at the worst disasters in terms of loss of human life.

Ten Deadliest Disasters (2000-2019)

	Disaster	Location	Year	Deaths	
	Earthquake & Tsunami	Indian Ocean	2004	226,408	
	Earthquake	Haiti	2010	222,570	
	Storm	Myanmar	2008	138,366	
	Earthquake	China	2008	87,476	
	Earthquake	Pakistan	2005	73,338	
	Heatwave	Europe	2003	72,210	
	Heatwave	Russia	2010	55,736	
	Earthquake	Iran	2003	26,716	
	Earthquake	India	2001	20,005	
	Drought	Somalia	2010	20,000	

This chart is dominated by earthquakes and tsunamis, which I do not think anyone would attribute to Climate Change.

Finally, this chart shows number of people affected.

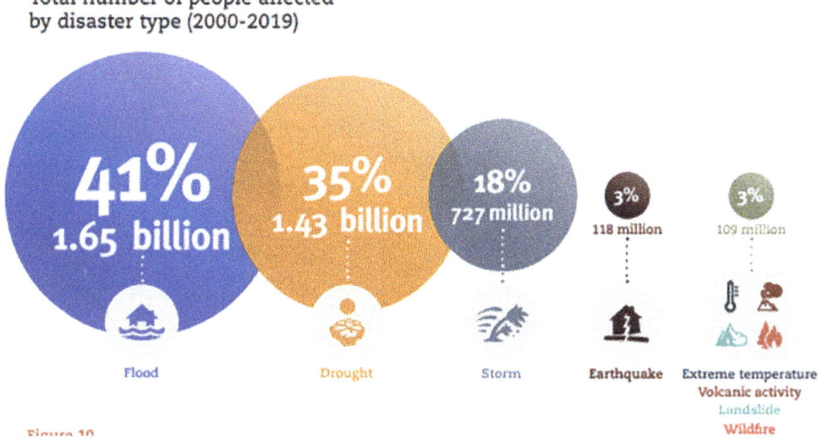

Figure 10

Here we see that floods and droughts dominate.

I cannot see how this affects any of the conclusions that I have drawn in previous sections of this chapter. This report is a good accounting of events, but the alarmist conclusion given in the introduction is not backed up by any connections between the events and Climate Change.

How do we then explain the following chart from that report, which is the one that made the News?

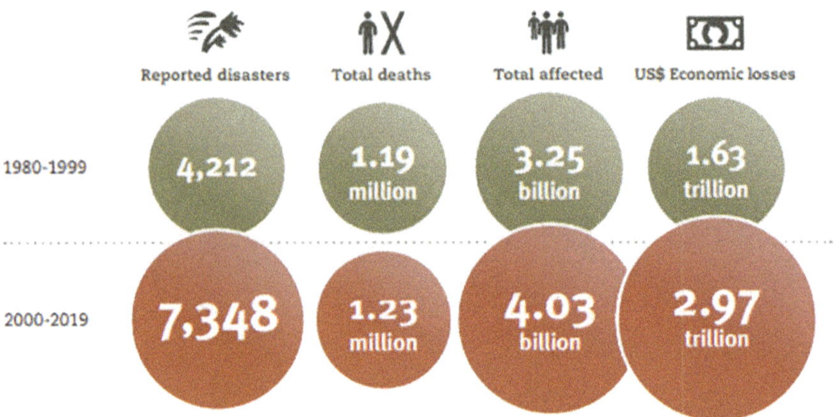

The number of people affected by disasters rose from 3.25 Billion to 4.03 Billion between the 2 test periods. This is an increase of 24%. That seems impressive, but not when you consider that the world population grew by 30% in the same period! (I used population from the median years 1990 and 2010).

The number of deaths was stable, indicating that per capita death rate has greatly reduced. Incredibly, this report claims that 511,000 deaths were "climate-related".

The number of disasters is shown to have increased, but it is strange that in the 20-year detailed trend chart from 2000 to 2019 there is no increasing trend at all. Was there a sudden jump in 2000 and then it was maintained? Unlikely.

The issue here is not the level of alarmism contained in this report, because that can be readily found in many publications. The issue is that this was issued by a United Nations body.

RISKS TO WILDLIFE CAUSED BY HABITAT CHANGE

Here I choose to interrogate the data for a few particular examples. These have all seen quite a bit of attention in the media and in the research community. They are coral reefs, insects, and polar bears.

1) Coral Reefs

The Great Barrier Reef (GBR), located along the coastline of Queensland in Australia, is one of the wonders of the world. Rich in marine life, much of it supported in some way by the corals that bloom in glorious colors, it is the world's largest single structure made by living organisms. Corals are the exoskeletons of tiny polyps, which are symbiotic with algae in the ocean.

Corals exist only in the tropics, from –30 to +30 latitude. The minimum ocean temperature where corals thrive is around 16°C. There is also a maximum, dependent on the species, and this is what may be causing distress in coral reefs.

It is reported that more than half of the Great Barrier Reef has disappeared since 1985, and this is usually attributed to human activity. Climate change is universally reported as the main root cause. There are two climate aspects relevant to this discussion: warming of the ocean waters, and acidification of the water due to increased absorption of CO_2. When temperatures rise, the algae die or depart, leaving behind the white exoskeletons of the corals. This is referred to as coral bleaching. Bleaching events can cause death of the corals, but in some cases, they recover with reduced reproductive capabilities for some period of time. Acidification is caused by a chemical reaction between water and CO_2, which produces carbonate ions that inhibit calcification. This has large implications not only for corals, but all shellfish in the ocean, which use calcium to build their shells. We will address acidification in Chapter 3's future world, because it does not appear to be a driver right now.

Ocean water surface temperature has risen in the GBR at a rate similar to what we have seen as the word-wide average, about .6°C since the 1950's. (T, 2016)

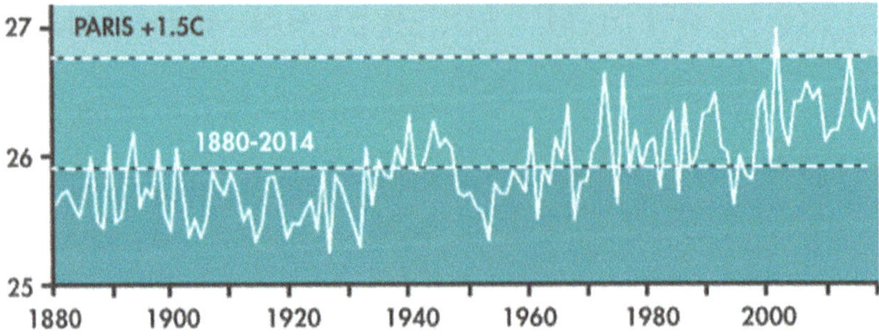

This chart shows average yearly temperature, but what is perhaps more important is the seasonal change, which drives the temperatures in summer above a tipping point where the algae are stressed. There are 3 drivers of temperature increase: global warming, normal seasonal fluctuation, and irregular seasonal fluctuation driven by El Nino events. El Nino is an irregular but recurring warming of waters in the tropical Pacific Ocean.

As with other phenomena described in this book, I have tried to put current conditions into historical context, in order to determine if global warming is a true culprit, as opposed to a silent witness to normal recurring weather patterns. Below is a chart taken from a comprehensive meta-study of corals going back more than 100 years. (J. M. Lough, 2018)

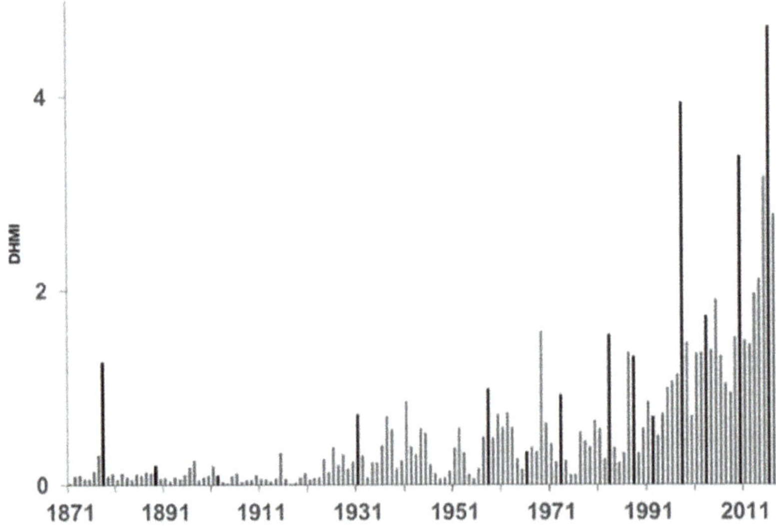

Average annual degree heating month index (DHMI) for 100 coral reef locations, 1871–2017 (grey bars). Black bars mark years with Niño 3.4 index ≥1.0 °C.

The study examined 100 coral reef regions world-wide, and this chart uses the Degree Heating Month Index (DHMI) to describe the progression of ocean temperature as it affects corals. Essentially, a 1°C increase in temperature lasting for one summer month would result in a change of 1 in the index. Each bar describes a calendar year, and it is the average increase over all 100 reefs. The dark black bars show El Nino years.

This is a clear upward trend occurring in the time period of global warming. It remains only to confirm that both high temperature and its duration are true causes of distress to corals. This is proved indirectly by the fact that mass bleaching events occurred during the years of highest DHMI. Mass bleaching occurred in 1998, 2002, 2005, 2010 and 2015–2016, corresponding well with years of high HMDI.

Below is shown the relationship between the basic Sea Surface Temperature (SST) and HMDI

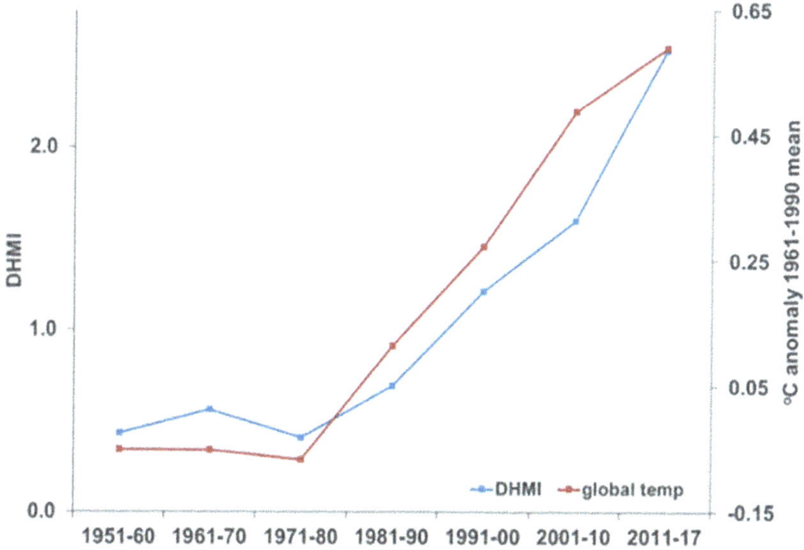

Average degree heating month index DHMI (blue) and global land and sea temperature (red) for 10-year periods, 1951–1960 through 2001-2010 and most recent 8-year period, 2011-2017.

I show this chart to illustrate how a seemingly small change in ocean temperature (.6°C) can strongly affect corals in particular, which are in distress only at peak temperature periods, well represented by HMDI (2.5)

Still, there are reefs in the world that are faring much better than the GBR. The Tubbataha Reef in the Philippines are located around a set of uninhabited islands 130 kilometers from any population centers. There is no fresh water source. It is a World Heritage site and is well protected, remaining relatively pristine. Some coral bleaching was observed in the world-wide mass bleaching of 2015-2016, but they have mostly recovered. Sea temperatures here are actually slightly higher than on the GBR, with summer temperatures reaching 30ºC.

With a lack of persistent human presence, and the relatively stable state of corals, the Tubbataha Reef might indicate that there are other mechanisms contributing to the decline of coral reefs around the world. In Australia, the reefs are shoreline for a 424,000 square kilometer "catchment area" that is used mainly for agriculture, and several rivers flow from this land into the ocean.

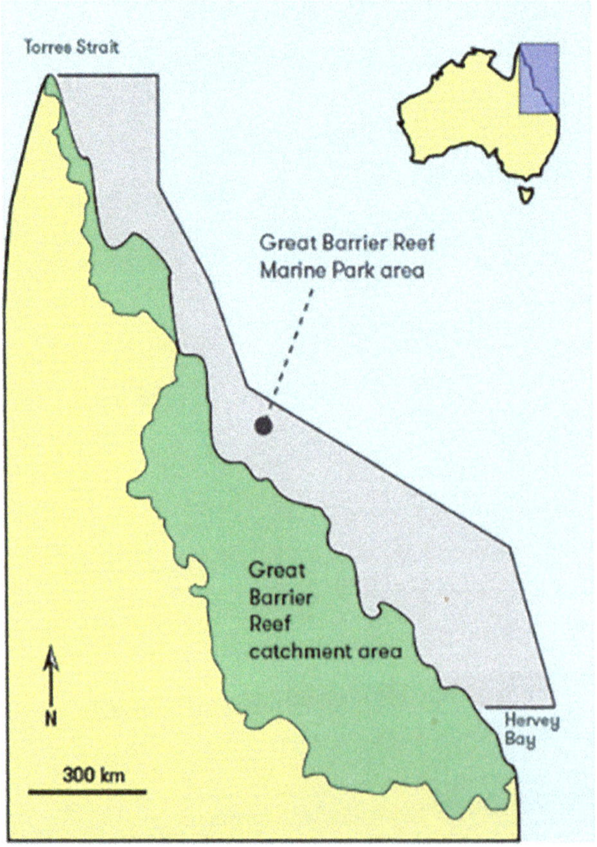

Below is an estimate of the outflows, taken from the Australian Academy of Sciences Website.

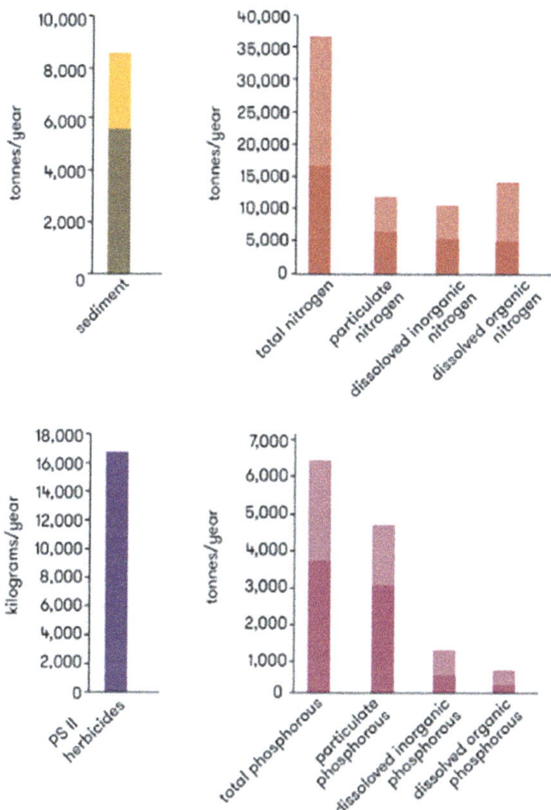

Sediment is made up of fine particles that can settle onto the corals, inhibiting their activity. The darker portions of the bars representing nitrogen and phosphorous originate in fertilizer used on the farms. These are thought to result in an exaggerated population of crown-of-thorns starfish which feed on the corals. Evidently, Herbicides are probably also a threat.

None of these factors are present in the Tubbataha Reef, and it appears to be healthier, sustaining itself better than the GBR. If it is true that 50% of the GBR is already gone, it seems unlikely that global warming is the sole culprit, and other actions need to be taken to protect it.

2) Insects

The contributions of insects to human well-being are probably underappreciated. The most evident contribution is the pollination of food crops by honeybees and other pollen-carrying species. Like the small plankton in the oceans, land-based insects are part of the food chain foundation in the animal world, and they play pivotal roles in many ecosystems in processing the decay of plant and animal matter.

In 2008 a paper was published on the effects of warming on insects. It has been referenced frequently by other scholarly works. The paper attempts to predict the long-term effects on insect populations up to the end of the century. We will discuss this part in Chapter 3. Here I reproduce the results of laboratory-level experiments on insects that were used to determine sensitivity to temperature. (Curtis A. Deutsch, 2008)

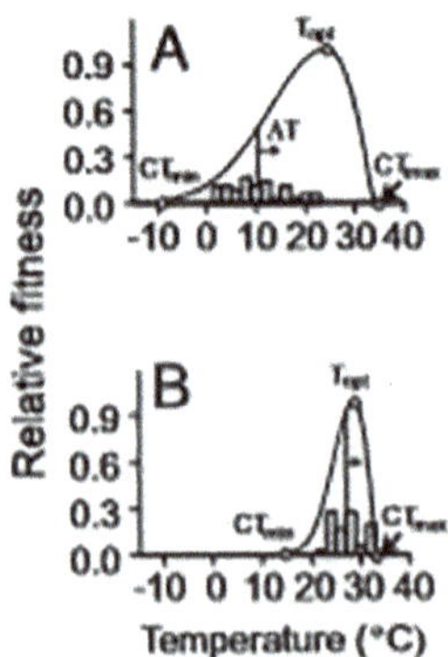

The graph identified as "A" applies to insect species in temperate climates, and "B" to those in tropical climates. The parameter "Fitness" is a scale the researchers developed to indicate the species' ability to function and reproduce in its environment.

A major conclusion to be drawn is that those species that already experience large seasonal temperature variations (at high latitudes) are well-adapted to change (Chart A), while those species that live in year-round high temperatures (low latitudes) are much less adaptable (Chart B). An interesting supplementary finding was that the same held true for insectivores, the small animals that feed on them.

I am reminded of the plight of coral reefs, which are existing only in tropical climates and are already experiencing distress with only small changes in temperature.

In one study (Garcia, 2018) made in the Luquillo National Forest in Puerto Rico, a case is made that recent warming has already had a severe detrimental effect on insect populations, and on insectivores. This forest is a protected region in Puerto Rico since a long time, so it is expected that other human interventions are not contributing factors. This is a tropical region that fits the profile exhibited in Figure "B" above. In 2011 – 2013 sweeps were made of the soil and also in the tree canopy to collect insect biomass and compare the results to similar sweeps made in 1976-1977.

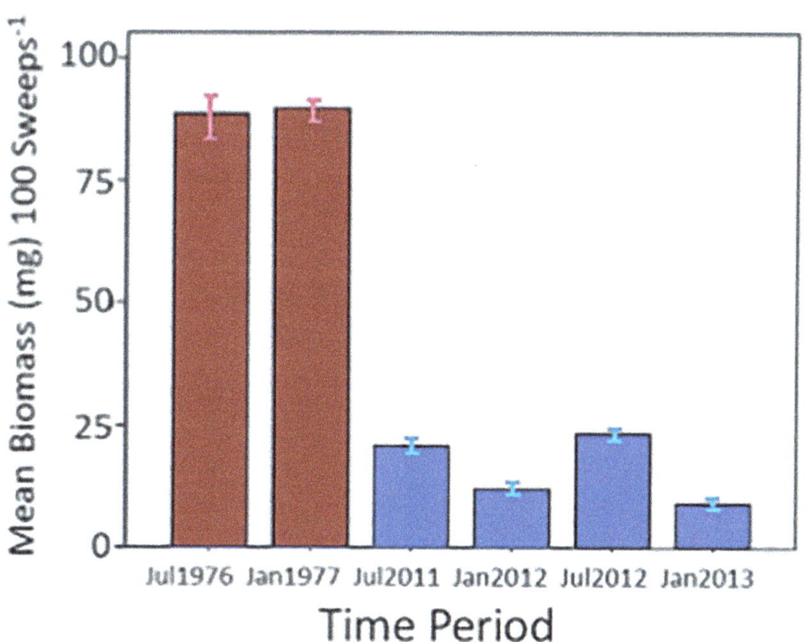

(Lister)

The results were certainly dramatic. Only around 20% of the biomass was detected. Similarly, there was an observed reduction in populations of insectivores.

The study attributes these results entirely to warming of the region. It is claimed that a temperature change of 2°C has been observed between the two study periods. Below, El Verde and Bisely are two local weather stations.

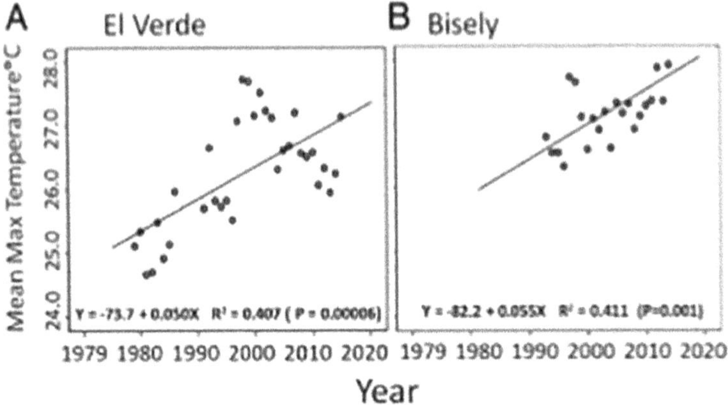

Can it be that 80% of all the insects are gone already from this forest? The authors claim of a 2°C change is questionable. The graph shows peak daytime temperature, not average temperature. Also, does the trend over many years really matter, compared to the temperatures at the times the studies were done? There appear to be some years with high temperature in early 2000's. Given the short reproductive cycle time of insects, would the temperature in 2000 matter when they are being investigated in 2012? I suppose it would matter if many species were made extinct in those few hot years, but that seems not to be the case, because in the paper the same species are discussed for the two different study periods. Therefore, I would claim that the temperature change is only 1°C, consistent with the rest of the world.

No other explanation is discovered for the decline in population. It is re-iterated in the conclusions that no other human interventions are present due to the protected status of the forest. However, if a temperature change of only 1°C is sufficient to reduce insect biomass by 80%, then there is indeed trouble ahead. Looking at the graphs from the Deutsch study, we see tropical insect populations are tolerant in a temperature range of about 25°C to 31°C *average*. The arthropods of the Luquillo forest appear intolerant to daytime *peak* temperature of only 27°C. This seems questionable to me.

An even less convincing case (Peter Soroye, 2020) is made for the decline of honeybees in a study that looked at historical weather patterns and bee populations. The premise can be summarized as follows: weather has changed, bees are in decline, therefore weather is responsible. Other studies point out alternative powerful factors implicated: pesticides, herbicides, invasive mites from "foreign" bee species, habitat destruction and poor land management. The European Union has banned some neonicotinoids, a class of pesticides, as these are thought to be a major contributor to the problem. In the USA and Canada, they are still widely in use. It is controversial because it appears to take time to have effect, therefore not obvious as a culprit. Bee deaths are mostly observed over winter inactive periods, recently suffering as much as 40 % loss.

Bees are present in all climates and on every continent except Antarctica. They live in regions where they hibernate in freezing winter months and then thrive in hot summers. It is difficult to imagine that a temperature change of only 1°C would have such disastrous effects on their numbers. They fall mostly into Figure "A" from the Deutsch study, living in and adapted to large seasonal temperature variations in temperate climates.

For bees, the jury is still out, but I suspect climate change is not implicated so far.

3) Polar Bears

The common narrative surrounding polar bears is the loss of hunting grounds caused by the retreat of Arctic Sea Ice. Seals are their primary food source, and they are most easily captured on the ice or on its edges. Seals hunt for fish and then go onto the ice to rest or take sun. Alternately, bears make holes in shallow parts of the ice and use them to find seals. It has been assumed that reduction in this habitat would have a catastrophic effect on bear population.

In 2017 National Geographic released a video showing a polar bear in distress, clearly thinned out, hungry and possibly ill. The caption read: "This is What Climate Change Looks Like". This video was perhaps the single most important factor that brought climate change to the forefront for much of the world's population.

The photographer who took the video later admitted that the reason for the bear's distress was entirely unknown. Perhaps loss of hunting habitat was implicated, but we will never know.

A good source for some statistics related to the conditions of polar bear populations is the Polar Bear Specialist Group (PBSG). I reprint the complete 2019 version of their table of data below.

Subpopulation	Subpopulation size			Subpopulation trend		Sea ice metrics 1979-2018		Human-caused removals 2013/2014-2017/2018	
	Estimate and uncertainty	Method and type of evidence	Year and citation	Long term (approx 3 generations)	Short term (approx 1 generation)	Change in date of spring ice retreat / fall ice advance (days per decade)	Change in summer sea ice area (percent change per decade)	5-year mean	
								Quota (bears per year)	Actual (% of total population)
Arctic Basin	Unknown			Data deficient	Data deficient	-9.8/14.2	-7.4	N/A	
Baffin Bay	2826 2059-3593	Genetic C-R	2012-2013	Data deficient	Data deficient	-6.1/3.9	-16.0	127.8	126.8 (4.8%)
Barents Sea	2644 1899-3592	Distance sampling	2004	Data deficient	Likely stable (2004 to 2015)	-16.0/25.3	-19.6	N/A	
Chukchi Sea	2937 1552-2944	Physical C-R with density extrapolation	2016	Data deficient	Likely stable (2000 to 2016)	-6.2/6.8	-31.9	58 (changed to 85 in July 2018)	15.4 (0.5%) in U.S. + approx. 33 (1.1%) in Russia
Davis Strait	2158 1833-2542	Physical C-R	2007	Data deficient	Likely stable (2007 to 2016)	-5.6/6.7	-16.3	QC + 75.6 (NU & GL)	64.4 (3.6%)
East Greenland	Unknown			Data deficient	Data deficient	-7.0/6.4	-7.9	N/A	
Foxe Basin	2585 2096-3180	Mark-recapture distance sampling	2009-10	Stable (1994 to 2010)	Stable (1994 to 2010)	-4.5/4.6	-14.1	QC + 117.3 (NU)	102.8 (4.0%)
Gulf of Boothia	1592 870-2314	Physical C-R	2000	Likely stable (2000 to 2017)	Likely stable (2000 to 2017)	-5.0/6.6	-12.6	72.4	61.0 (3.8%)
Kane Basin	357 221-493	Genetic C-R	2013-2014	Data deficient	Likely increased (1997 to 2014)	-6.9/4.6	-0.8	11.0	3.0 (2.2%)
Kara Sea	Unknown			Data deficient	Data deficient	-9.7/9.9	-22.7	N/A	
Lancaster Sound	2541 1759-3323	Physical C-R	1995-1997	Data deficient	Data deficient	-5.6/3.8	-7.4	84.9	80.5 (3.2%)
Laptev Sea	Unknown			Data deficient	Data deficient	-6.7/7.2	-17.5	N/A	
M'Clintock Channel	284 166-402	Physical C-R	2000	Very likely increased (2000 to 2016)	Very likely increased (2000 to 2016)	-4.2/4.7	-9.0	8.0	7.0 (2.7%)
Northern Beaufort Sea	980 825-1135	Physical C-R	2006	Likely decreased (2006 to 2018)	Likely decreased (2012 to 2018)	-7.2/3.6	-6.5	142.5 (SB+NB)	83 (SB+NB)
Norwegian Bay	203 115-291	Physical C-R	1997	Data deficient	Data deficient	-1.8/2.9	-1.7	4.0	2.0 (1.0%)
Southern Beaufort Sea	907 548-1270	Physical C-R	2010	Likely decreased (1986 to 2010)	Likely decreased (2001 to 2015)	-9.7/6.6	-35.3	134.4 (SB+NB)	63.4 (3.3%)
Southern Hudson Bay	780 590-1029	Mark-recapture distance sampling	2016	Very likely decreased (1986 to 2016)	Likely decreased (2012 to 2018)	-1.8/3.1	-8.5	51.6	56.4 (4.7%)
Viscount Melville Sound	161 93-329	Physical C-R	1992	Data deficient	Data deficient	-5.5/7.1	-5.4	7.0	3.6 (2.2%)
Western Hudson Bay	842 562-1121	Mark-recapture distance sampling	2016	Very likely decreased (1993 to 2016)	Likely decreased (2011 to 2016)	-5.9/3.3	-21.8	51.6	24.6 (3.5%)

Given the landscape, and the fact that polar bears are relatively solitary creatures, it is acknowledged that tracking population is difficult. However, using this best source, it can be seen that sub-populations are either stable, increasing, or declining, or insufficient data is available. It is generally accepted, however, that the species is not in distress at this time.

Neither does there appear to be any correlation between population trends and loss of summer sea ice, which is included in the table. Intuitively I find this not surprising. Bears are intelligent animals and should therefore be able to adapt. Hunting seals on sea ice is not their only source of food. Many northern human populations can certainly attest to the fact that they are also excellent scavengers.

I also wonder about the seals. They are in the ocean to hunt fish, and they will continue to do so. If the ice is missing, they will surely return to the rocky shores from which they came. Would the bears not find ways to hunt them there?

The last 2 columns in the table shows data on "human removals". This means hunting. Around 800 bears a year are shot, legally, by indigenous people, or by hunters guided by indigenous people. A quota system is maintained so as not to overstress the population. In a sad way, this is an indication that polar bear populations are healthy. I wonder, though, what reaction National Geographic would get if they put a video on YouTube showing a healthy bear being shot and stripped of its fur.

RISING SEA LEVELS CAUSED BY MELTING OF GLACIAL ICE AND WATER THERMAL EXPANSION

THE ARCTIC

To understand warming effects in the Arctic, it is important to look at ancient history where long periods of both warming and cooling have occurred. There are many studies done where ancient weather patterns are discerned from "proxy" data, meaning data in lieu of direct measurements. Here I refer to one particular 2010 study by G.H. Miller et al, "Temperature and Precipitation History of the Arctic". Building on the work of other researchers, and also adding more data and insight, it is a remarkable read. It humbles me in my efforts here to explain the science to a more general population.

Over exceptionally long periods of time, in the millions of years, the Earth's climate has experienced greater extremes than anything we are seeing today. Among the many drivers are variations in the Earth's seasonal wobble, extreme volcanism, changes in solar irradiance and extreme variations in greenhouse gases. Sixty-Five million years ago, around the time of the dinosaur extinction, the Earth was so warm that the Greenland Ice Sheet did not even exist. Today, it is 3.2 kms thick! This is why today dinosaur bones can be found even in Arctic regions.

Examples of the proxy data used to determine temperature and precipitation are: Distribution of fossil pollen, position of northern tree line, dendroclimatology (tree rings), oxygen isotope content of calcareous shells, lake sediment and ice core water, and the fossil presence of specific planktonic species that are known to live in particular thermal conditions. It must certainly be a monumental task to assemble all of this evidence extracted from disparate regions to arrive at mean temperatures and precipitation levels. Confidence is gained by comparing conclusions on the same data from various different researchers and models. Below is one conclusion, a graph of northern hemisphere temperature covering the last 2000 years, with the results of different researchers superimposed. (G.H. Miller, 2010)

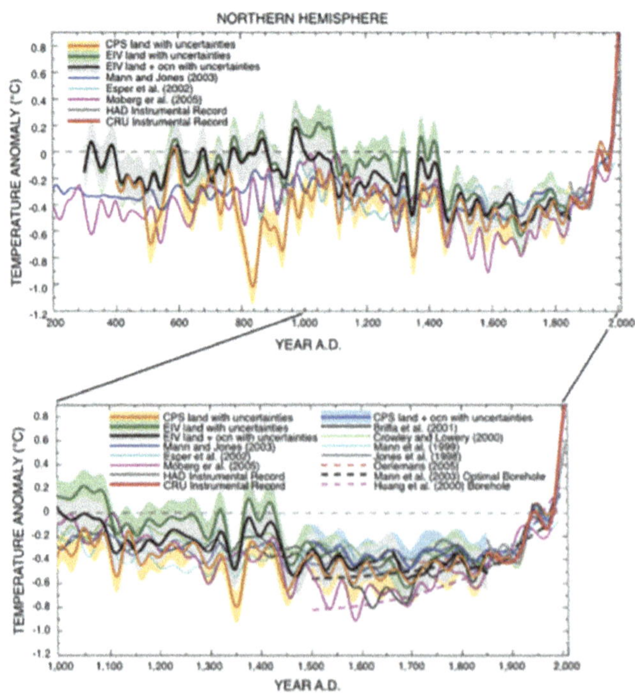

The reference temperature is from 1950, prior to the current anthropogenic warming period. All data is proxy except for the recent period of direct measurement shown in red from the mid 1880's onward. It can be clearly seen that recent temperature rise is unprecedented in the last 1000 years.

Temperature increase in the arctic has significantly exceeded the global average. Below are 3 charts showing the whole Earth, and the trend in the Arctic only (north of 60° latitude), and the Arctic Ocean temperature. These have been obtained from Berkeley Earth, with data provided by the UK Hadley Center.

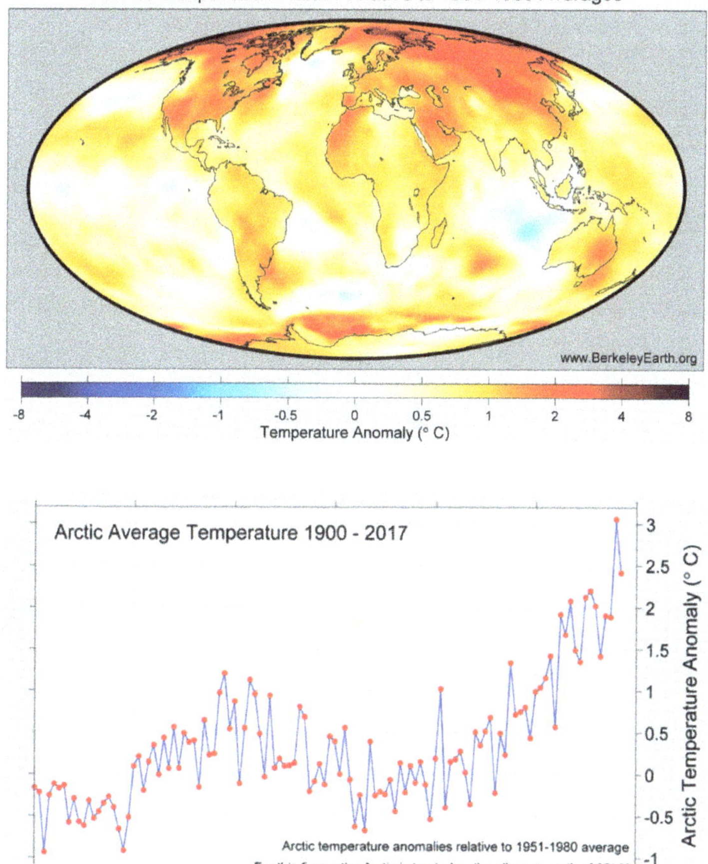

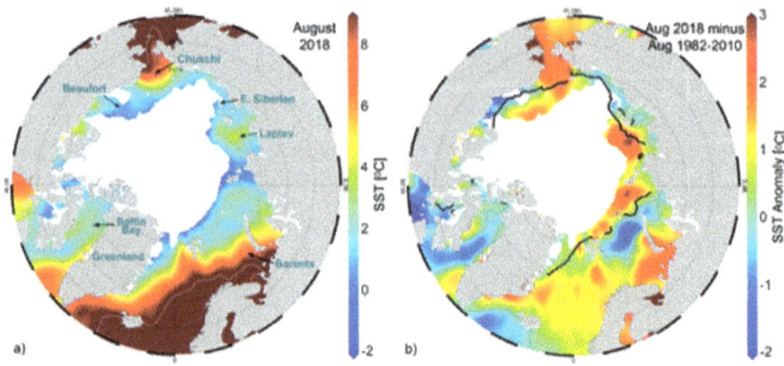

It can be seen that the beginning of the rise occurred around the year 1980, same as elsewhere on the planet. This magnitude of change is not seen near the other pole, in the Antarctic.

The common explanation for this enhanced warming is the loss of sea ice in summer months. The ice is highly reflective, so it tends to send solar radiation back to space. As air temperature warms, more ice melts, exposing the ocean underneath. The ocean is close to being a black body, meaning it absorbs the sun's radiative heat, and then sends it back to the atmosphere. This creates more warming, and more melting of ice.

The third image shows the 2018 ocean temperature in August (when sea ice is at minimum), and the change in temperature (anomaly) compared to the average from 1982 to 2010. Grey areas are land masses, white is ice, and colors represent Arctic Ocean temperature. Clearly the ocean has seen warming as a result of this greater summer exposure to solar radiation. The anomaly is much higher than the global ocean average. The very warm region shown in the Barents Sea is mostly caused by influx of very warm water from the Atlantic Ocean.

I wondered how the Arctic can get so much warmer than the rest of the world, considering that the air circulates. Typically, it takes only about 2 weeks for a mass of air to circulate around the globe, although there can be considerable variation. Therefore, why does the temperature anomaly not even out? The answer seems to be that the air temperature in the Arctic is intricately linked to the temperature of the ocean water, and the water does not circulate nearly as quickly.

Is the recent loss of sea ice in summer months unprecedented, or is this part of some natural cycle? Since 1979 the extent of the ice has been measured very accurately by satellites. Prior to that, there are many sources on information, such as dedicated flyovers and shipping logs. All this historical data has been summarized, producing the following chart. (Fetterer, 2016)

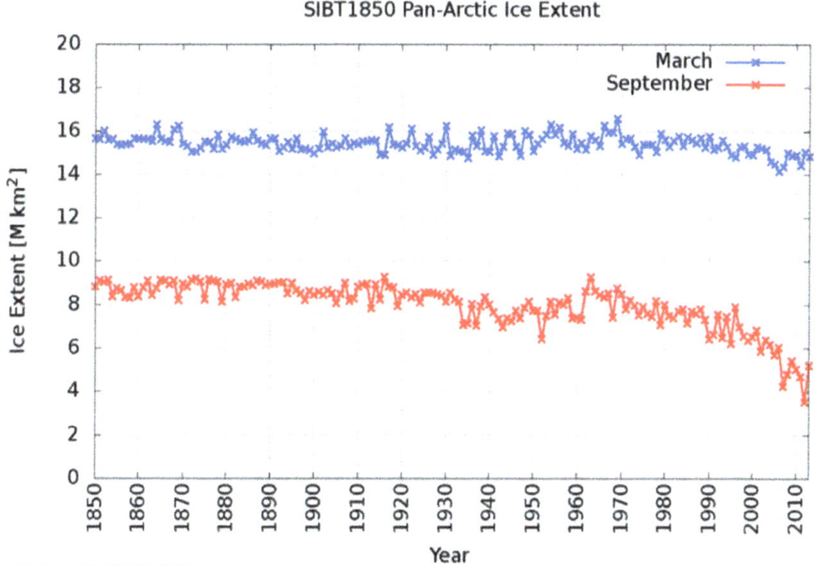

The decline of winter ice in March is so far only marginal, perhaps an encouraging sign if the ice can regenerate itself. The decline in summer is dramatic, and indeed it is unprecedented in recent history.

At first glance this could look like a positive development. Humans living in the arctic will mostly be happy with warmer temperatures. Summer shipping lanes are opened up, bringing more commerce to the arctic. However, there are downsides. Warming of ocean waters contributes more to global warming. That extra heat we see in the arctic atmosphere must eventually migrate south because all air on the planet is mobile. Essentially, the sea ice acts as a planet cooler as it reflects solar radiation. As previously discussed, polar bears and other marine mammals can be negatively affected by a change in habitat from its historical state.

There is another effect of warming temperatures, perhaps the most important, that has been the subject of much attention from both popular media and the scientific community. Warming of the Arctic and Antarctic is causing land-locked ice to melt, leading to rise in sea level all over the world. Melting of sea ice has no effect on sea level because it is already included in the total volume of water. However, historically land-locked ice will add to the total volume if it melts. There has always been a certain amount of melting in summer months, which has then been offset by the addition of snow and ice in the winter, keeping sea level relatively constant. The recent warming in the arctic has been extreme enough for the volume of melt in summer to be much greater than what is replaced in winter months.

Historical trend in sea level rise is given below.

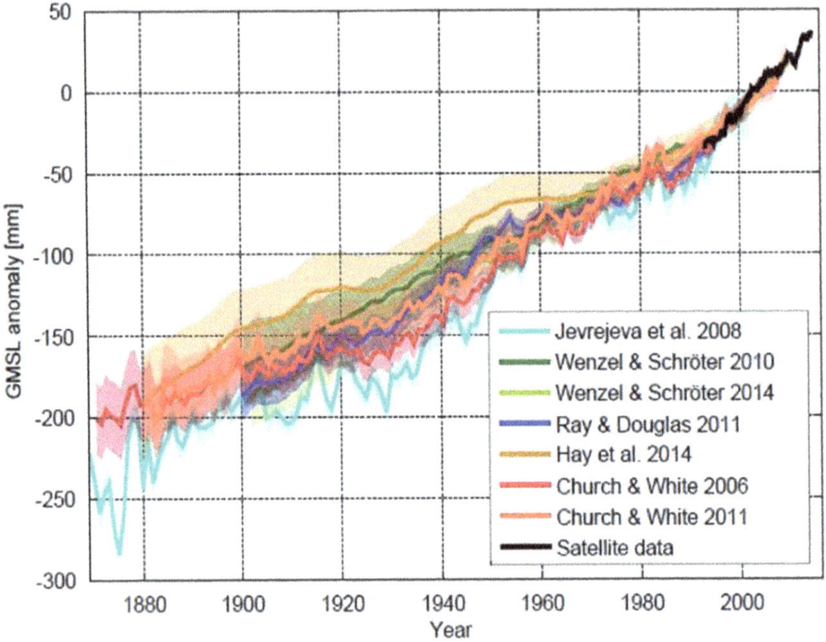

In this graph, there are multiple curves representing different interpretations of historical tidal gauge data by different researchers. Notice that in the recent satellite era the data all come together.

Interestingly, the rising trend starts in 1880, at the beginning of the period of record, and there is an acceleration in the years of global warming (roughly, 1980 onwards). The rate of rise (the slope of the curve at any point) has recently reached about 3 mm/year, compared to an average rate of 1.5 mm/year over the years 1880 to 1980.

With respect to the effect on sea level caused by warming, there are two factors to consider. Firstly, the change in ocean temperature causes a thermal expansion of the water itself, causing it to occupy more volume. Secondly, the addition of water from melting of historically landlocked ice will increase the ocean's mass and volume.

Recall from chapter one the trend of ocean temperature, reprinted below.

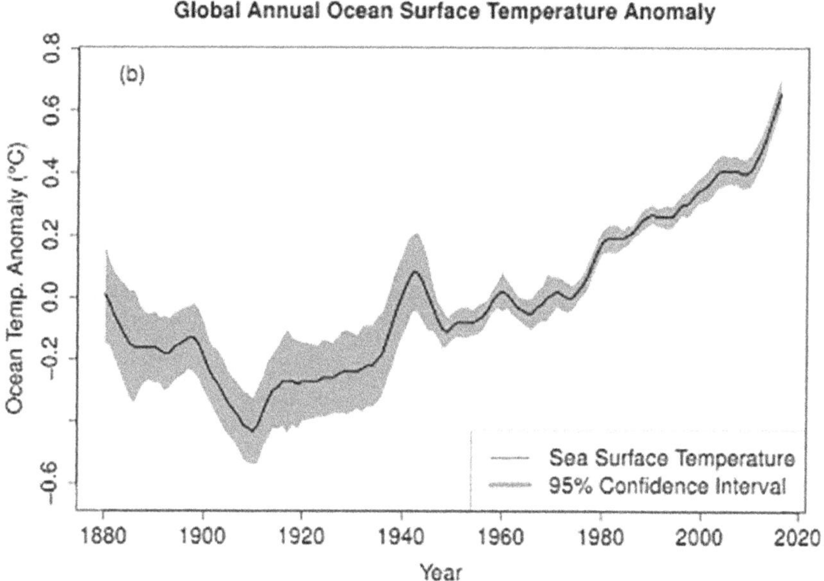

We see a falling trend from 1880 to 1910 (as also seen in the 2000-year graph above), then a rising one all the way to present day. This is somewhat out of sync with the sea level trend in the early years, indicating that warming of the water and the associated thermal expansion was not in play. The reason for continuous rise of sea level since the mid-1880's is highly likely the melting of glaciers on land in the North, and in mountainous regions at lower latitudes. There was a so-called "little ice age" (LIA) from the 16th to the mid-nineteenth century, when glaciers were growing, with annual winter snowfall exceeding summer melting. The LIA is thought to have been caused by high volcanic activity, which causes particulate matter in the atmosphere to block incoming solar radiation. The effects can be seen in the global temperature graph plotted on a 2000-year scale in Chapter 1. Since then, the world has been gradually warming and the glaciers have been gradually receding. The period of anthropogenic warming started later and has accelerated the retreat of the glaciers.

A dramatic example of glacier retreat is the Athabaska in Alberta, Canada. This is a frequently visited site because it is accessible with a walk of only a few hundred meters from the Icefields Parkway.

The tip of the field as seen in this photo has retreated by 200 meters since 1992. More importantly, the thickness or depth has reduced so much that it will probably not survive the next 100 years regardless of what happens with our weather.

Glaciers like the Athabaska store water, which is partly, and gradually, released in summer months, feeding downstream lakes and rivers. This flow is a guardian against drought in hot months. Disappearance of such glaciers therefore impose a risk for humans.

Important contributors to future sea level rise are the large landlocked ice shelfs in Greenland and Antarctica. Below is an image from NASA on the estimated net loss of ice since 2002 from Greenland.

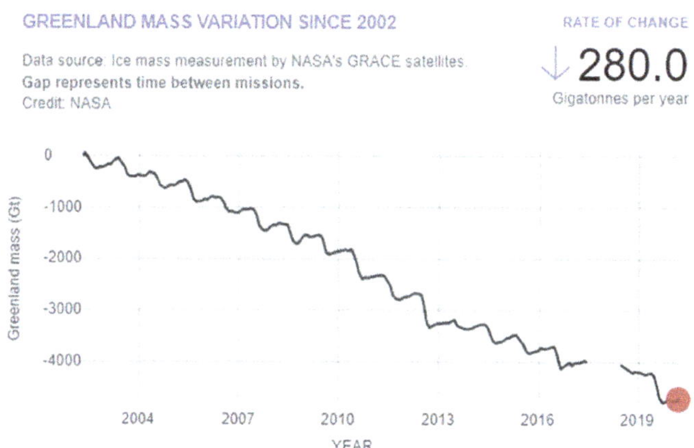

Units are in Gigatonnes (Billions of tonnes, where 1 tonne = 1000 kg). A further image below shows the rate of loss of ice, and especially the locations, which are all around the edges of Greenland. The edges are where the warm water is in partial contact with the ice, and also where local air temperatures are less affected by the massive volume of the ice itself.

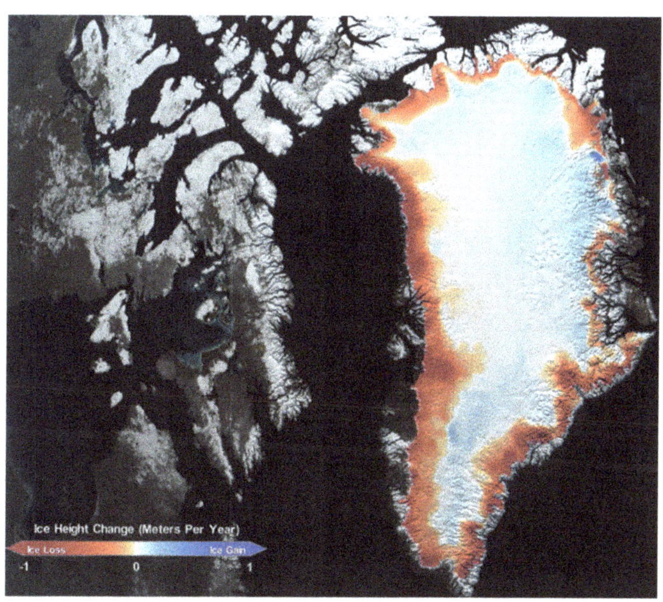

This data from NASA is relatively recent and very sophisticated. It uses satellite imagery to measure the altitudes of the ice sheet surfaces, and it includes effects of the enormous weight of the ice on its position. However, as with other phenomena in this book, I look for long term trends to ensure that recent events are not just a short-term anomaly. For Greenland, I found one painstaking study which is highly informative on a longer scale, going back to 1972. (Jérémie Mouginot, 2018)

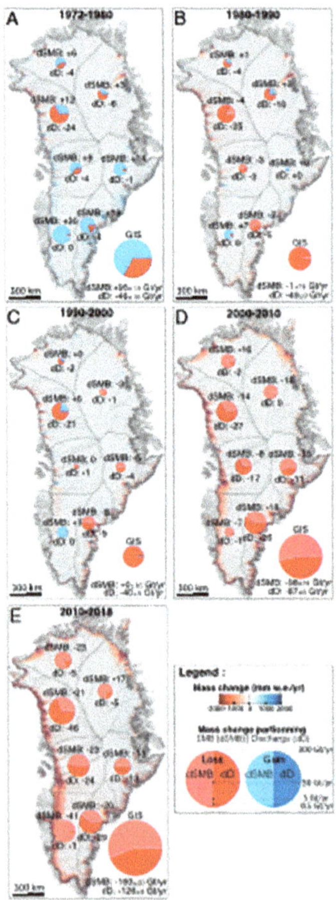

This chart shows the progression of ice loss over the decades. Red and pink colors show loss by either melting or discharge of ice into the ocean, and blue shows mass gain. The size of the circles in each of the regions indicates the amount of mass loss or gained. The acceleration of mass loss is readily apparent. Plotted another way below, the change in slope, or acceleration since year 2000 is also apparent.

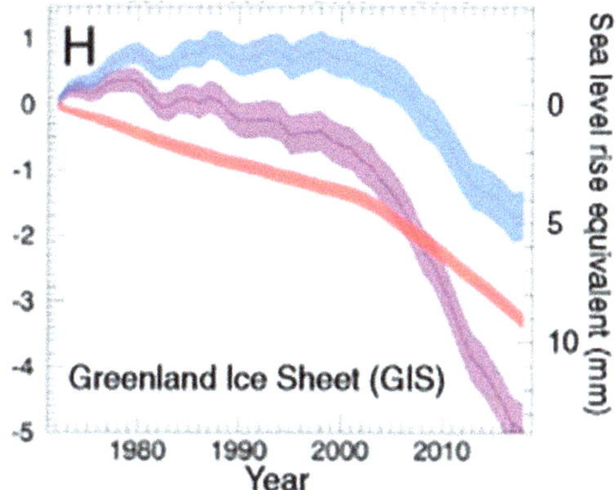

In blue is the mass of ice added (by snow) or subtracted (by melting). Red is the amount of ice lost by direct ice discharge into the ocean, and purple is the net mass loss or gain. Also included here is the estimated sea level rise corresponding to the ice mass loss.

ANALYTICAL STUDY ON LOSS OF SUMMER SEA ICE

All of this is part of the common narrative on warming of the Arctic and the associated rise of sea level. I still had some nagging doubts about how all this is playing out. Intuitively, I know that radiation from the sun is a far more powerful influence on ice melt than simple convection from surrounding air that is warmed by greenhouse gases. Recall also from Chapter 1 that solar radiation has seen no important upward trend in recent history. Therefore, I needed to convince myself by doing a kind of heat balance that independently explains the accelerating cycle of summer ice melt.

Firstly, I will describe an experiment that takes place every spring around my home. This is one experiment that is not likely to be reported in a scholarly article any time soon, but I think it demonstrates well the source of my confusion. I live in Montreal which is close to 45° latitude. I have a duplex, or, in more international jargon, a row house. It is connected on both sides to other row houses, so there are no sides exposed to the outdoors. I have only the front and back yards. Every winter we get large amounts of snow. In the front, all of the snow that needs to be shoveled away winds up in a substantial mound in the restricted space between houses. An impressive mountain is created. In the back, the yard is unused all winter. There is no shoveling away of snow, so it remains in a flat rectangular shape exactly where it falls.

In the spring, it could be expected that the mound in front should take much longer to melt than the flat area in back. From a heat transfer point of view, the mound has an extremely high ratio of volume to exposed surface area, so heat transferred by the warming air should take much longer to take effect. In fact, the exact opposite happens. The front mound melts completely 2 to 3 weeks earlier than the snow in back. This is because the front is exposed to direct sunshine, while the back is entirely shaded by the house. This demonstrates that radiation is much more powerful than air convection, even though it is only active for a small part of the day, and even though the snow has a high degree of reflectivity.

Let us look at some numbers. I choose the month of June and a latitude of 65°N to do the heat balance in the Arctic. I focus on melting of sea ice, not the landlocked ice sheets. Firstly, here are some images that give us some important inputs to the calculations:

Air Temperature (Left) and Sea Surface Temperature (Right) North of 60°N

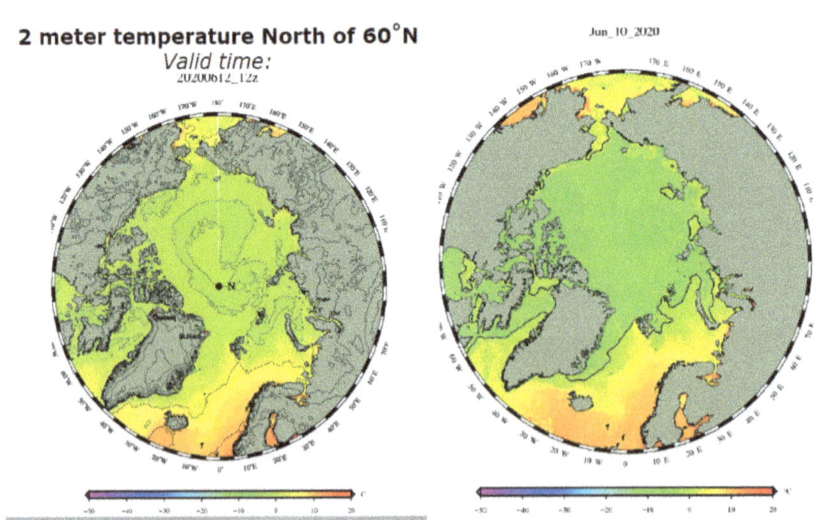

These images from the Danish Meteorological Institute (DMI) show the air and sea temperatures for a day in June in 2020, with the same color scale used for both. Grey areas are land masses, colored areas are either sea ice or ocean water. It can be seen that the ocean, ice and air temperature are all similar. The ocean cannot have a temperature lower than about -2°C, otherwise it would be ice. The ice cannot be much colder than the ocean that it is in direct contact with. The air is also hovering around 0°C due to its proximity to the ice or ocean. The air temperature over permanent ice shows that it can never exceed 0°C due to the local effect of the ice itself, as in this DMI image covering the permanent ice region north of 80°N.

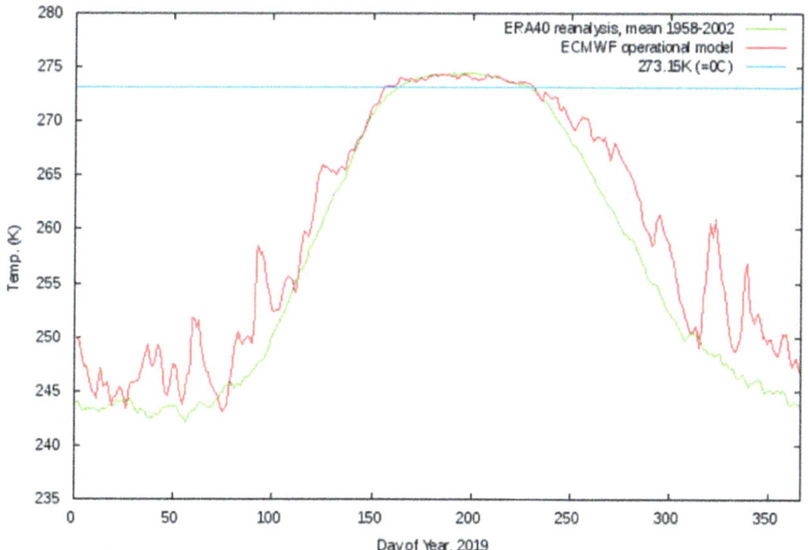

In summer months this "clipping" of the temperature keeps it at around 0°C over ice. Note also from this image that the warming in recent years, represented by the difference between the red and green curves, occurs mostly in non-summer months! This is caused by residual heat in the ocean water that has been warmed during the summer.

From a heat transfer point of view, I will make the following assumptions: Ice is at -2°C, Ocean and air are at 0°C. This will represent the status prior to the years of anthropogenic warming. For current status, I will assume that the air has increased by 2°C, and the water by

1°C, in line with the known temperature anomalies. For radiation, I will assume an increase in the amount of radiant heat reflected back to the ice caused by an increase of CO_2 from 210 ppm to 420 ppm. For this I take data from a paper (Harde, 2017) where the reflection coefficients are calculated to be .6181 and .6216, respectively. Other assumptions are given in the table below for clear sky daytime.

Radiation		Convection, Air to Ice	
HEAT LOAD SUN TO ICE			
Sun to Earth, Outer Atmosphere	1362	Assumed Average Wind Speed (m/s)	3
Absorption by Atmosphere	362	Convective Heat Transfer Coefficient h (W/m^{2}C)	7.1
Heat Load at Equator	1000	Pre-Industrial Temperature Difference Between Air and Ice ΔT (C)	2
Reduction from Scattering at 65°N Latitude	689	Pre-Industrial Convective Heat Load Onto Ice = hΔT	14.2
Albedo of Ice (Fraction Reflected)	0.6		
Heat Load onto Ice Q1	275.6	Current Temperature Difference Between Air and Ice ΔT (C)	4
		Current Convective Heat Load Onto Ice = hΔT	28.4
HEAT RADIATED BY ICE TO ATMOSPHERE AND SPACE		**Conduction, Ocean Water to Ice**	
Emmissivity of Ice ε	0.98	Conductivity of Ice k (W/mC)	2.23
Temperature of Ice T (K)	273	Thickness of Ice in Summer t (m)	2
Stefan-Boltzman Constant v (Wm^{-2}K^{-4})	5.67E-08	Pre-Industrial Temperature Difference Between Water and Ice ΔT (C)	2
Radiation from Ice = ε v T^4	308.6	Pre-Industrial Conductive Heat Load onto the Ice = kΔT/t	2.23
Pre-Industrial Fraction Reflected Back at CO_2 of 210 ppm	0.6181		
Pre-Industrial Net Radiation Out From Ice Q2	117.9	Current Temperature Difference Between Water and Ice	3
Current Fraction Reflected Back at CO_2 of 420 ppm	0.6216	Current Conductive Heat Load onto the Ice = kΔT/t	3.345
Current Net Radiation Out From Ice Q2'	116.8	**Final Heat Balance**	
		Radiation + Convection + Conduction	
Pre-Industrial Net Radialtion Heat Load Q1 - Q2	157.7	Pre-Industrial (W/m^2)	174.2
Current Net Radialtion Heat Load Q1 - Q2'	158.8	Current (W/m^2)	190.6

This study shows an important daytime increase in heat load onto the sea ice, even though, as expected, the unchanged radiation from the sun is the prime driver.

I extended this study to include cloudy sky and nighttime variables, Clouds are highly variable. Albedo of clouds can be as low as .1 (high thin clouds) to .9 (low thick clouds). I have made the crude assumption that average albedo is .5, and I use this factor both as reflector for high frequency solar radiation, and reflector for low frequency outward radiation from the ice. Here is the result (positive indicated melting, negative indicates freezing).

Era	Pre-Industrial	Current	% Change
Clear Sky Day	174.2	190.6	9.4
Cloudy Sky Day	95.4	111.3	16.6
Clear Sky Night	-101.4	-85.0	-16.2
Cloudy Sky Night	-42.5	-26.7	-37.3
ASSUME 75% DAYLIGHT, 25% NIGHT, 50% CLEAR, 50% Cloud			
Averages	83.1	99.2	19.4

I find that for the long days of June at 65°N latitude, there is a 19% increase in heat load onto the sea ice. Reduced amounts would be expected in the months leading up to and following June.

The assumptions used here are crude but respect basic fundamentals. I conclude that the accelerated loss of summer sea ice can indeed be attributed to anthropogenic warming.

What does this mean for land-locked ice that is the main contributor of sea level rise? There are important differences:

1. There is no contact with ocean water. In practice, there can be some contact at the base and this can lead to sliding of ice sheets into the ocean, but we will ignore this for now. The result is that we remove conduction from ocean water from the equation.

2. The initial ice temperature is much lower. Since the ice is not in contact with the warm ocean, it is exposed only to the very frigid winter air temperatures and can therefore start the summer with extremely low temperature as compared to sea ice. For this study, we must still assume the surface of the ice is at 0oC, because it cannot melt otherwise. However, we should keep in mind that melting may initiate later in the summer season.

3. The outer edges of the tall ice sheets of Greenland can be attacked by convection and radiation on two sides, that is, from the top and from the side, see below.

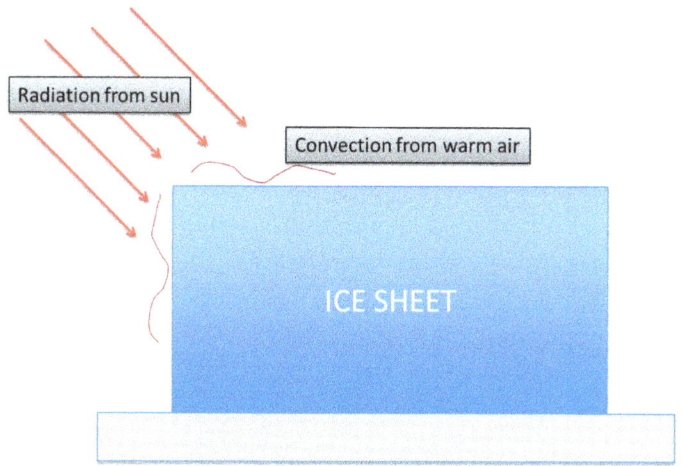

If we assumed that both radiation and convection effects are doubled on the edges, and conduction from the ocean is removed, there is of course a higher heat load than what is calculated for the sea ice, but the percentage change compared to pre-industrial times is not changed by much.

It is clear from observations that only the edges of the ice sheets are melting, while the middle is not affected. This is because the land-locked ice is much colder than sea ice. It requires this double-attack onto the edges to create any melting at all. The centers of the ice sheets are just too cold for two reasons: the thickness, up to 2 km in Greenland, creates a giant thermal barrier to melting; and the surface temperature at the end of winter is very low because, unlike the sea ice, it has no contact with warm ocean water.

THE ANTARCTIC

Now let us look briefly at Antarctica and the South Pole. Antarctica is a large land mass covered mostly with permanent ice, much like Greenland. Unlike in the Arctic, temperatures in Antarctica there have not increased much. This is because the Arctic is basically a large ocean with relatively warm water, while Antarctica is ice locked onto land. In the ocean around Antarctica, there is also significant loss of sea ice in the summer, approaching 100%. However, this has always been the case even before the global warming period, so our climate is already adapted to it.

Air temperatures near the ocean, such as at the US McMurdo Station, reach around 0°C in summer and fall to -30°C in winter. In the interior, such as at the Russian Vostok Station, the corresponding values are -30°C and -60°C. Warmest of all is the Antarctica Peninsula, which is a thin strip of land surrounded by warm ocean water.

Here is a record of surface temperature made by proxy records (ice cores, black line), compared to hard instrumental record (grey line), going back to 1800. (David P. Schneider, 2006)

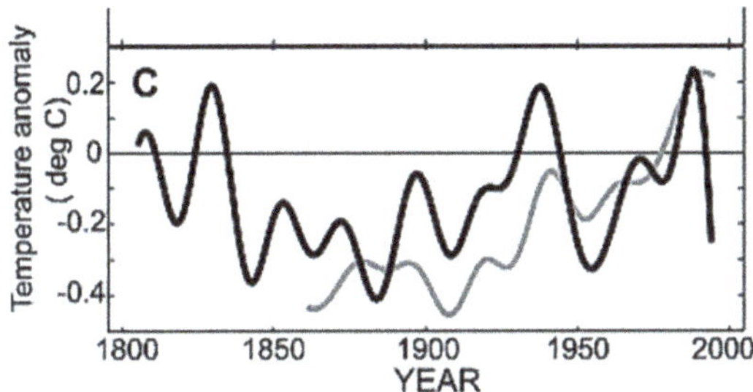

The temperature anomaly is with respect to the average from the years 1961 to 1990. It can be seen that the anomaly is exceedingly small over many decades and is not increasing up to the end of the record in 1990.

For more recent data, here is the record up to 2010 (BROMWICH, 2014)

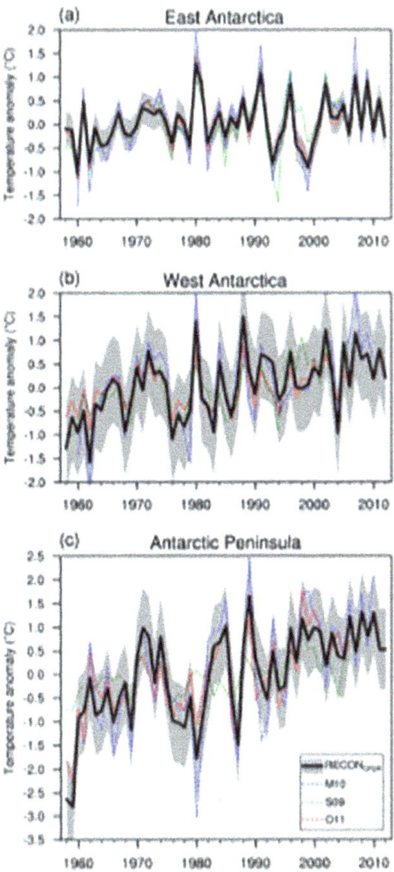

In this case the anomaly is with respect to the average of the years 1960 to 2006.

The authors conclude that there is a warming trend in West Antarctica and the Peninsula, while no warming is observed in the East. It is odd, however, that warming seems to have started at the beginning of the record in 1958, and the rate of warming has not changed in the period after 1980.

In spite of only a small change in average air temperature, there is still a significant amount of summer melting of land-locked ice.

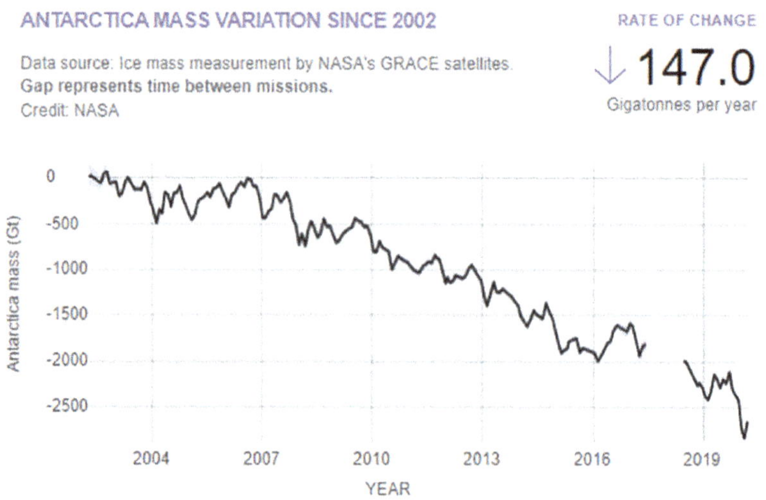

Below we see the regions affected.

There are regions losing ice but also regions gaining ice over the time period illustrated, from 2002 to now. The regions on the left, known as the Amundsen-Bellingshausen Seas (ABS) and the Weddel Sea (surrounding the Peninsula) are where most of the ice loss occurs.

In a 2013 paper (Megha Maheshwari, 2013) a comprehensive review of the Sea Surface Temperature (SST) was made for the entire Southern Ocean. Below are 2 images, the first showing the temperature anomaly for the Ocean over time, as compared to the 1982 – 2010 average, the second showing the temperature difference between the two extreme years.

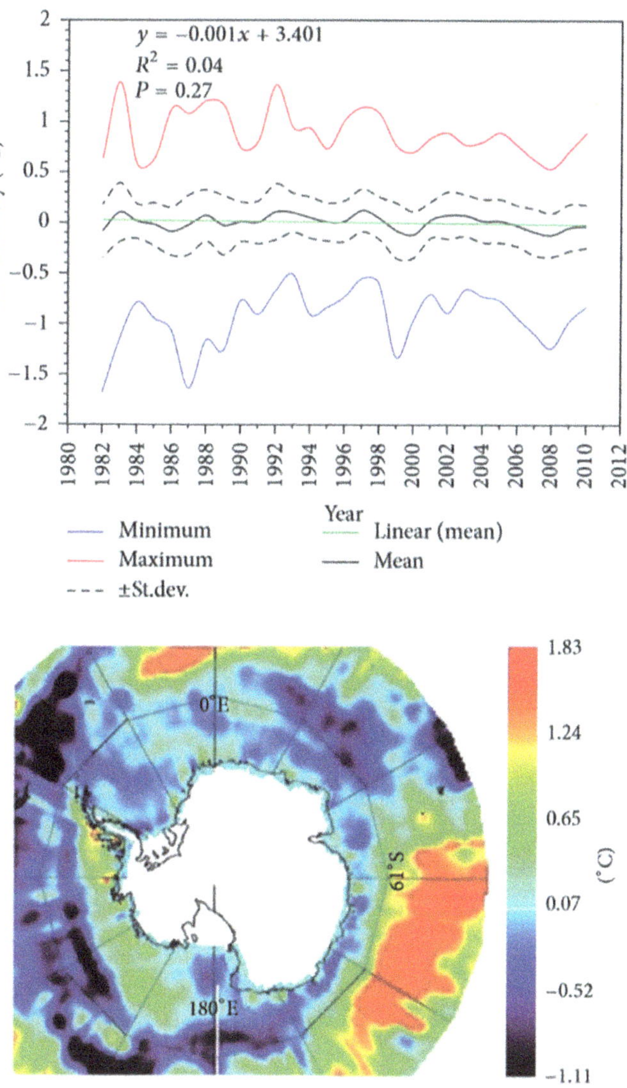

It can be seen that the net temperature anomaly for the entire ocean is zero, with certain regions becoming cooler or warmer. In the 2nd image, the SST around the Antarctic continent shows a positive anomaly only at the ABS and Weddel Sea, and its magnitude is not more than .6°C, except around the Antarctica Peninsula where it is around 1°C. This peninsula contains the Larsen Ice Shelves which have seen some dramatic partial collapse events since the mid-1990's.

It is evident both in Greenland and in Antarctica that melting of the ice that serves to increase global sea level occurs on the edges of the ice sheets. The enormous central cores of the ice probably cannot be affected until it is attacked from the sides, as discussed in the section on the Arctic. We will revisit this in Chapter 3.

We can conclude that warming in Antarctica, and the melting of land-locked ice, is less severe than in the Arctic. The total amount of ice melt from all sources is not yet large enough to affect coastal areas inhabited by humans, but it is certainly a harbinger of things to come.

CLIMATE CHANGE IN THE PUBLIC EYE (PART 1)

There is a persistent media blitz regarding climate change. Almost every evening news show, every edition of a newspaper, every nature show, has references to the latest findings. Typically, the warnings get more and more dire. I perceive a feedback mechanism where enhanced public awareness tends to drive some researchers to seek out and identify more harbingers of doom in their work. Everyone needs to get on the bandwagon.

Here is a small example. In a popular documentary on Netflix, a segment was dedicated to Lake Tanganyika in Africa. It was declared that climate change there is causing fish stocks to be greatly reduced. Curious, I went looking for the research behind it. I found a paper in the journal Nature that used the following argument: water temperature has increased by .6°C, wind speeds have reduced significantly, and lake sediments show less productive organic activity over time. Therefore, climate change has had drastic effects on fish population. The change in wind speed and the temperature change are said to be the cause of reduced upswell of nutrients form deep in the lake. No explanation is given as to why wind speed is related to Global Warming. Some of the claims in the paper are challenged by two other subsequent papers published in the same journal. However, that did not make it into the Netflix show seen by millions of people.

Still curious, I decided to investigate what might be happening around Lake Tanganyika. There are four countries that border on the lake, and these countries have increased their population by 141% since 1980. The fish in the lake is an important source of food for the local population, so the demands on the lake have greatly increased. To discourage overfishing, there is an international governing body that sets rules, The Lake Tanganyika Authority, and local authorities are meant to police them.

In surveys of fishing activity done in 1995 and 2011, the number of fishermen and number of vessels have more than doubled, the number of motorized boats has tripled, and the number of gill nets have increased by a factor of 8. Illegal beach seines (shoreline dredging of the ocean) have increased by more than 50%. Clearly, if we need to know the source of the reduction in fish stocks, look no further.

The decline in bee populations which is discussed earlier in this chapter is another example. With all the stress we put on bees, primarily in applying insecticides to the very crops we expect them to pollinate, why assign blame to an amount of warming that is a only a tiny fraction of the temperature range to which the species is demonstrably well adapted?

A National Geographic article declares the following: "First Mammal Species Recognized as Extinct due to Climate Change". The animal is a Bramble Cay melomys, a small rodent that lived on a tiny island near the Australian Great Barrier Reef. The island maxes out at 10 feet above sea level. Any inundation of water drastically reduces the land area and therefore the rodent's habitat. The article cites the rise of sea level of 8 inches since 1901 as the major cause. However, as we saw above, the majority of that rise is caused by the exit from the Little Ice age, which in turn was caused by volcanic activity, not anthropogenic warming.

Another headline I came across declared that Australian summers are now 4 months long, instead of 3! The following logic is used: if you take the pre-industrial average temperature on the first and last days of summer, you will find that with today's warming, these values are reached 2 weeks earlier at the beginning, and 2 weeks later at the end of summer. So, it feels like summer is a month longer. This works out because Australia is on average a relatively temperate climate, meaning that changes from summer to winter are not very great. If you extend the logic to the equator, where the annual temperature variation is only about 3°C, then you find the new summer is 6 months long! At the north pole, it is longer only by a few days. What does all this mean? Nothing. The temperature change is still 1°C in Australia, the same as everywhere else. However, the headline says summer is 33% longer. Somehow that must be 33% worse than before, I guess.

Examples like this abound. It can be argued that heightening the general public's awareness of climate change in this way is a positive outcome. Maybe so. From my point of view, however, it does us no good service to sweep under the Climate Change rug our many other shortcomings in the management of our environment. Finding the right root cause for each problem is the only way to get to solutions.

We will explore more of this in Chapter 6.

CHAPTER CONCLUSIONS

In general, the impact of global warming on human quality of life has so far been minimal. The same can be said for the effects on the environment and animal welfare, although there are harbingers of greater trouble in the future (Chapter 6).

The claims of increased severity of damaging storms and heavy precipitation is not verified. Records show that activity in the last 4 decades is in fact not more severe than before.

Wildfires in Australia have been severe but are still within historical norms. In California, there is a worsening trend, and it is clear that warming and drying of the climate has had an important impact. This might be the most serious physical warning that we have about the future.

Tropical coral reefs are already showing distress under warming ocean temperatures, although there is some evidence that other human activities may also be implicated.

Shrinking summer sea ice in the Arctic can be attributed to Global Warming, caused by the observed 2°C increase in air temperature and 1°C increase in ocean water temperature. The same can be said for local melting of land-locked ice in Greenland and Antarctica, but so far, the effects on sea level are manageable.

Dire warnings about climate change abound in the media and also in many scholarly articles. I find that many of these are not founded in good science, especially when effects in the current world are examined. Rather, there is a growing perceived obligation among both media and researchers to assign culpability to Climate Change when other simpler explanations may exist.

CHAPTER 3

CONSEQUENCES OF WARMING IN THE FUTURE WORLD

ONCE WE CONCLUDE THAT HUMAN-DRIVEN emissions are the cause of global warming, as we did in Chapter 1, then it is evident that the march to even higher temperatures is inevitable. Even if harmful emissions were to cease immediately, the world will continue to warm until natural decline of those aerosols occurs. The more likely scenario is that emissions will continue, and warming will accelerate.

In this Chapter I will refer often to data contained in the various reports of the Intergovernmental Panel on Climate Change (IPCC), at least in the areas of study that the Panel has undertaken. The IPCC is the internationally recognized source of data related to climate change. The IPCC does not do research on its own, rather it draws from, and often sponsors, qualified researchers in Academia and Industry.

Before presenting any projections, I need to define some terms. The IPCC has created four domains of the future which each correspond to a different level of emissions. These domains are called Representative Concentration Pathways (RPC's). RPC2.6 represents the most optimistic scenario, where the world converges on quick, coordinated strategies to bring emissions to near zero. RPC8.5 is at the other end of the spectrum, where emissions continue almost unabated by any corrective actions. RPC4.5 and RPC6.0 are intermediate levels.

I will frequently refer in this chapter to projected conditions in the year 2100, and, more often than not, the worst scenario RCP8.5. If 2100 seems too far away, we must consider the fact that greenhouse gas emissions take many decades to attenuate naturally once the emissions are terminated or greatly reduced. Any plan for year 2100 must start with mitigative actions long before then. I have chosen RCP8.5 as a benchmark because it describes a world where no real effective action is taking place, which is the prime direction at which we are currently pointed. It includes the assumption that emissions will continue to rise at historical levels of increase. This is why we see emissions in 2100 at around 100 Gt/yr, while in 2010 it was only 34.

Below is a chart that provides a "big picture" view of the future.

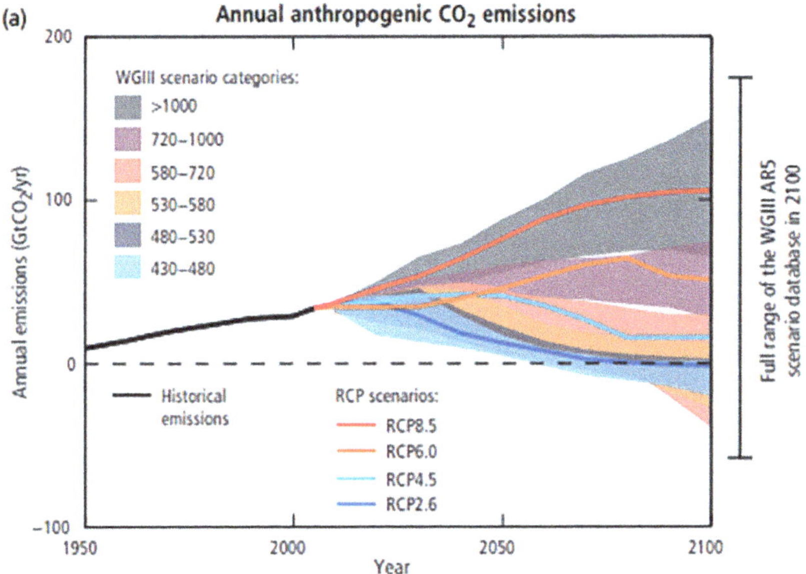

The solid lines show average projected emissions of CO_2 for the 4 RCP scenarios. The shaded areas give a band of probable atmospheric CO_2 content associated with the emissions. Similar charts exist for other GHG's. We see that the optimistic RPC2.6 scenario requires a near-zero emissions target by the year 2050.

It should be noted that all of the IPCC's projections do not yet include further increased CO_2 content caused by melting of northern permafrost, even though projected temperatures stipulate that such melting will occur. Permafrost contains frozen plant matter which, when allowed to decay, will naturally release CO_2.

Effects of warming on the carbon cycle is addressed by some but not all of the climate models being used by the IPCC. This is an area for enhancement in future modeling. The carbon cycle is the sum of natural phenomena that cause emission and absorption of CO_2. It is generally accepted that warming will cause further increases in natural CO_2 emissions.

Next, we see below a chart that predicts mean global temperature rise as compared to pre-industrial times, with the CO_2 emissions and atmospheric content as the variables.

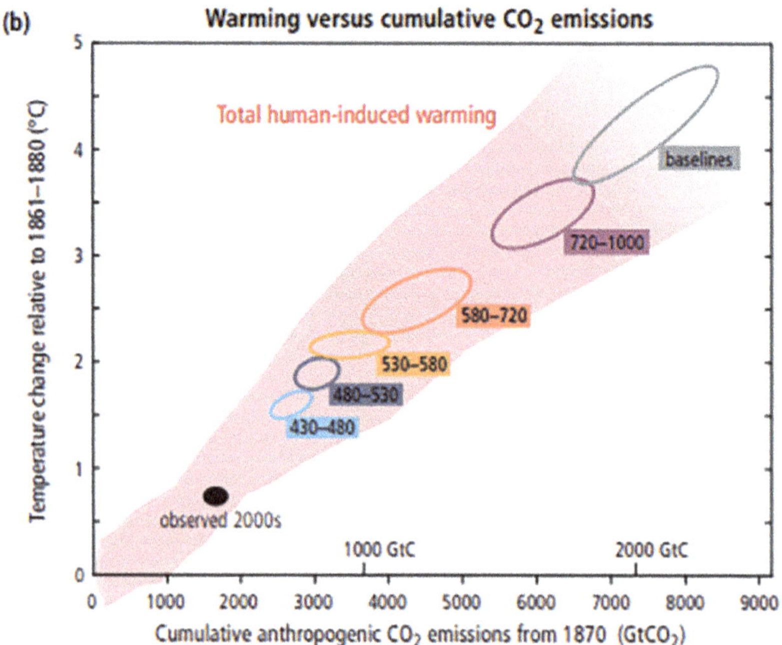

Finally, temperature predictions vs time for the 2 extreme scenarios.

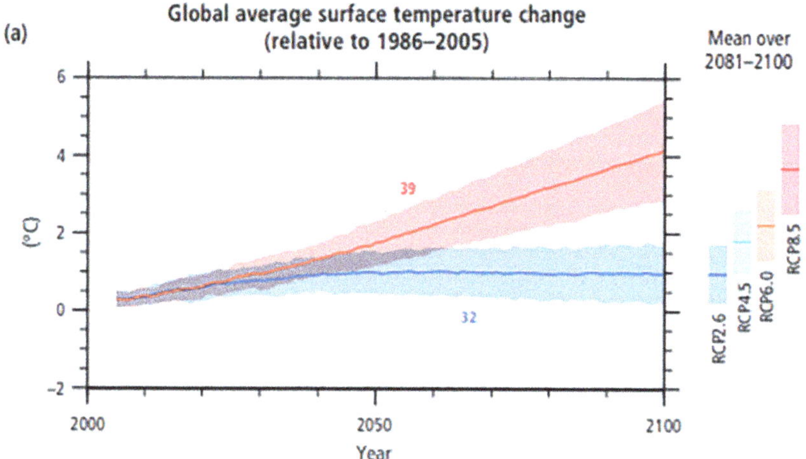

These are taken from The IPPC's fifth assessment reports, dated 2014. In 2021 the 6th assessment reports will appear with updates.

I note here that regional temperature differences are predicted, so that some areas would warm more than others. In particular, it is predicted that air temperatures over land will increase more than those over the oceans.

I have convinced myself by reading the material that this data is as robust as can be expected, and it is certainly derived from analytical processes that are far beyond my ability to reproduce. Buried inside these summary charts are detailed assessments of every source of CO_2. This is required because as humans produce more, there is always some being produced by natural causes, and also some being absorbed into the oceans and into plant life by photosynthesis. Scientists now have many decades of measured data with which their analytical models can be calibrated. I am also heartened by the tendency in these reports for the authors to express the degree of confidence in the accuracy, which is a critical part of the Scientific Method.

I include below one other figure that gives a comprehensive view of the effects of all of the different greenhouse gases (Julie Arblaster (Australia), 2013) by showing their individual contributions to radiative forcing onto the planet.

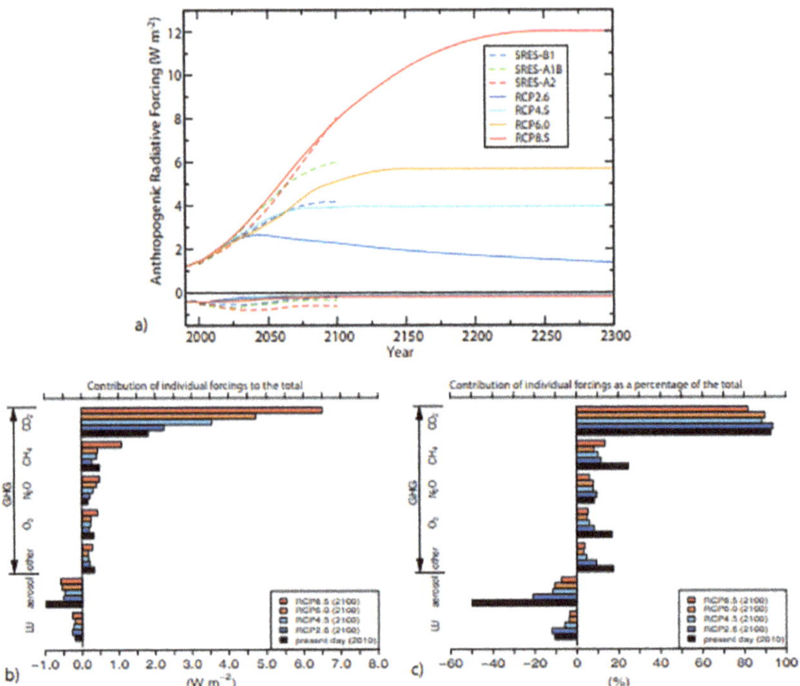

CO_2 is certainly the dominant contributor. Note that there are columns indicating negative forcing, meaning those items contribute to cooling of the planet. These are mainly aerosols, or particulate matter in the air that blocks out a portion of solar radiation. The majority of aerosols are naturally occurring, such as volcano and forest fire discharge. In these charts are included only the anthropogenic portion, meaning aerosols produced by human activity. The most important example would be pollutants produced by the burning of fossil fuels.

For the purposes of this book, the level of accuracy of the predictions of future global mean temperature is not particularly important, once we accept that warming will continue to occur. In fact, it is evident that the confidence band is very wide anyway, partly because the IPCC cannot predict the response of world leaders and governments to the cries for emissions reductions. Later in this book I will look in my crystal ball and try to predict such things. I would like to focus here on the processes used by scientists to predict the things resulting from Global Warming that matter most to people, which I believe are sea level rise, extreme weather as it affects humans, and extreme weather as it affects animals and plants.

SEA LEVEL RISE

In the Pleistocene era, at an average of about 4 million years ago, it is believed that atmospheric CO_2 was at about 400 ppm, like today. The world was 3°C warmer and the ice sheets in Greenland and Antarctica were considerably smaller, meaning also that sea levels were much higher. Conversely, in the last ice age, at around 20000 years ago, sea levels were considerably lower, as the great ice sheets accumulated more and more of the world's water. These facts are verified by methods similar to those already described in Chapter 2: measurements from cores drilled out of the ice sheets, and proxy methods using plant fossils and ocean floor sediment.

Today the Greenland ice sheet has a startling maximum thickness of around 3.2 km, but it is only several meters thick at its extreme edges where melting occurs. In Antarctica, the ice sheets have a maximum thickness of 4.8 km, and it holds an estimated 70% of the world's fresh water. In recent times the topography of these ice sheets is measured continuously by satellite imagery, so we have a good idea of the present status and the rates of change over time since the 1980's. As described in Chapter 2, the total sea level rise since the end of the little ice age in the mid-19th century is about 250 mm. It turns out that most of that has been caused not by melting of the great ice sheets, but by melting of glaciers in mountainous regions of the Earth. I remind the reader that part of sea level rise is also caused by thermal expansion of the warmed ocean water.

What follows are some charts from Chapter 13, "Sea Level Rise", of the IPCC's Fifth Assessment Report (John A. Church (Australia), 2014). The images (a) and (b) are all anomaly relative to the average of years 1986 – 2005.

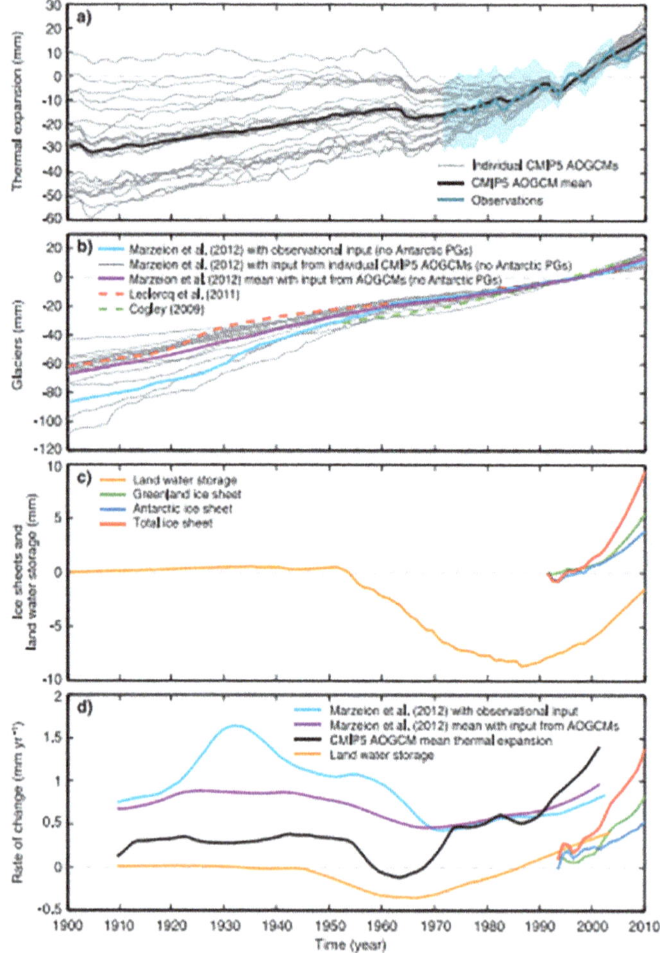

It can be seen from image (b) that glaciers have been the most significant contributor, and image (c) shows that only in very recent years have the ice sheets begun to have an impact on sea level.

In the images below, comparisons are made between observed sea level rise and the analytical models that attempt to simulate the observations.

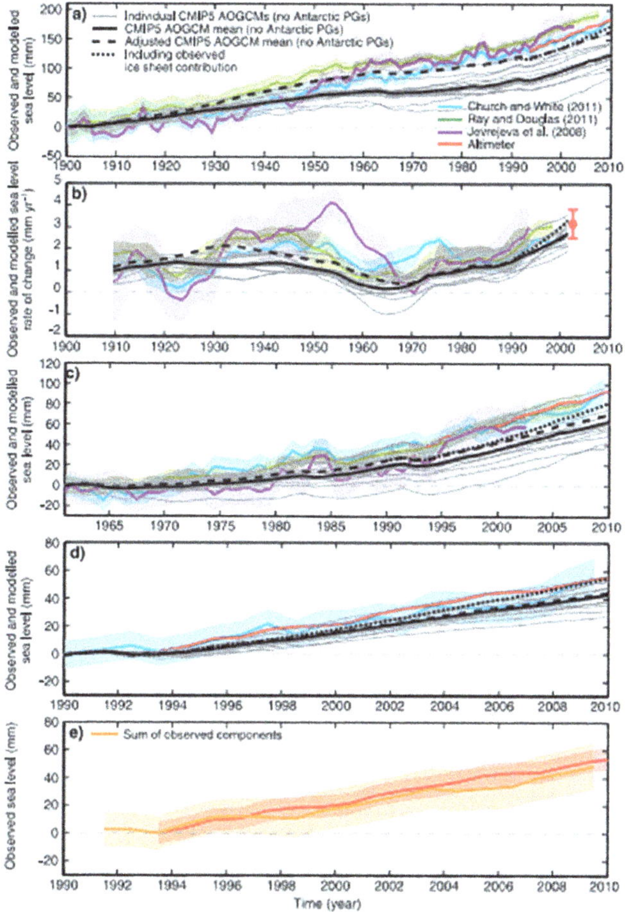

The fact that models are predicting quite well the observations gives some confidence that predictions of the future using these same models can be realistic. Before presenting that data, I will describe how these models work. They are called process-based models, meaning they attempt to make predictions with analytical tools. For example, prediction of ocean temperatures is made using all of the elements of heat transfer that are implicated: radiative forcing from the sun, outward radiation from the ocean, greenhouse effects of the atmosphere, cloud effects, and conduction of heat within the ocean water itself from surface to inner depths. For the glaciers and ice sheets, analytical models are made of the ice geometry using finite elements, and all of the heat transfer phenomena are again included. It is an iterative process where changes in the geometries and temperatures of the ice from previous iterations are re-inserted as a starting point for subsequent iterations. These models attempt to do what I did in chapter 2 when trying to see if global warming should be affecting the summer Arctic sea ice, except that these models are orders of magnitude greater in sophistication. I have faith in them to the extent that it is difficult to imagine a better way, but even the authors will claim that such complex phenomena cannot be predicted without significant error margin.

Other models exist which are referred to as semi-empirical models. The assumption is made that global mean temperature and sea surface temperature are direct indicators of sea level, so past observations of those few parameters are used to predict the future. These models are so far considered to be less accurate.

Below is the big conclusion. Solid lines are the mean predictions, and the shaded areas provide confidence bands.

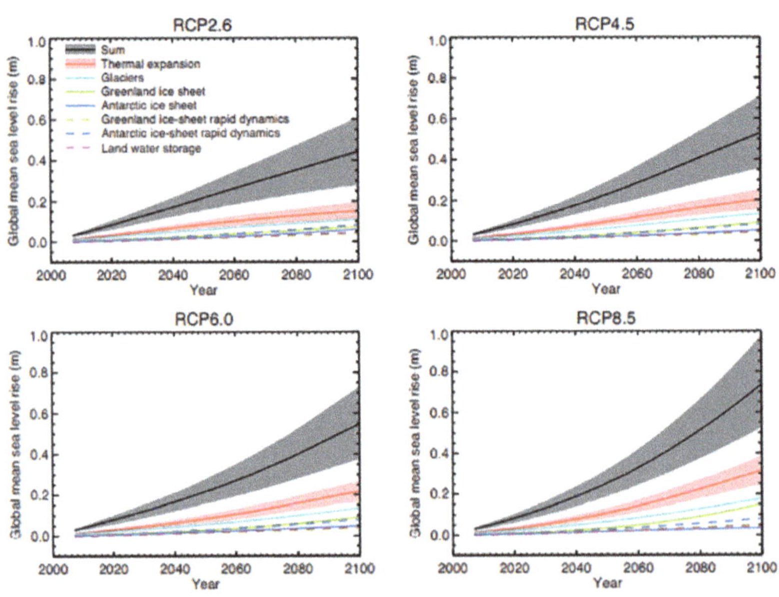

In the worst scenario RCP8.5, a rise of .7m is predicted, and in the best scenario RCP2.6 it is .4m.

It turns out that the expected most important factor in future sea level rise is the thermal expansion of the ocean water caused by its own warming. It is surprising that melting of the great ice sheets has such little effect in this calculation, because that is where all the media attention seems to be centered. Presumably, if we extend this projection beyond the year 2100, the ice sheets will have more impact.

These projections were published in 2014, and there has been considerable media attention since then that point to a worsening situation. I therefore looked for more recent projections. In 2017 a paper was published (Robert E. Kopp, 2017) that included physical modeling of ice shelf breakaway in Antarctica. In this model summer meltwater is used to provide mechanisms whereby large swaths of ice currently land-locked can slide away into the ocean, releasing much more water than previously predicted. It should be pointed out that previous models included the effect of moving ice sheets but were simplistic in that their velocities were assumed to be linear over time and were based on observations, not physical modeling. The reasons why this aspect is needed in the calculations has recently been well demonstrated by a huge break-away from the Spalte glacier in Greenland. The ice released from the glacier into the ocean measured 110 square kilometers.

The images below show projections for the conventional model identified as K14, and the new physical model identified as DP16. These charts show only the mean lines, although in all cases there is a considerable scatter or confidence band.

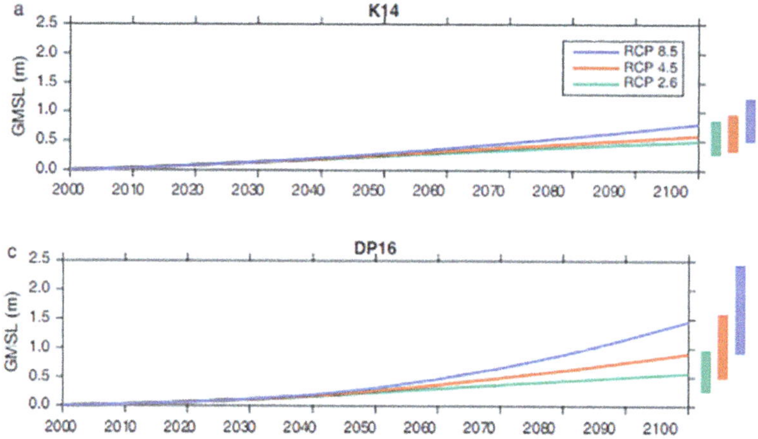

For scenario RCP8.5, the projection of sea level rise by 2100 goes from less than a meter to almost 1.5 meters, a significant difference.

There is more data like this out there, and in each case, there is some degree of uncertainty and disagreement. At this point, rather than trying to zero in on more and more accurate projections, I would rather go and look at the consequences of seal level rise in the future world.

Currently there are around 600 million people living along ocean coastlines. It is likely there will be some effects on all of them, but I show a chart below from the Kopp paper that quantifies populations that would be subject to direct inundation of water.

Population exposure (millions of people)					
Region	Total pop.	RCP 2.6/K14	RCP 2.6/DP16	RCP 8.5/K14	RCP 8.5/DP16
Current population occupying land exposed to inundation under 2100 RSL projections					
World	6836	94.3 (73.3–127.6)	97.4 (75.0–131.1)	108.2 (82.3–153.5)	152.5 (106.2–235.5)
China	1330	26.3 (19.1–37.5)	26.9 (19.8–38.3)	30.2 (21.8–45.0)	42.9 (28.3–67.0)
Bangladesh	156	7.6 (5.5–10.6)	8.0 (5.7–11.1)	8.9 (6.5–14.0)	14.0 (8.9–23.5)
India	1173	6.8 (5.2–9.1)	7.1 (5.3–9.4)	8.0 (5.8–11.4)	11.5 (8.0–18.5)
Indonesia	243	5.4 (3.8–8.0)	5.6 (4.0–8.4)	6.3 (4.6–9.9)	9.8 (6.1–16.6)
Vietnam	90	11.5 (9.1–15.1)	11.7 (9.3–15.5)	12.9 (10.1–17.0)	17.0 (12.7–23.9)

The last 2 columns show expectations for the RCP8.5 scenario currently considered by the IPCC and the revised estimate based on Kopp's new projections. Basically, between 100 and 150 million people would have to move inland, and overwhelmingly the effects occur in Asia. Indirect effects could be substantial.

Some rivers will also be affected, each in its own way. Rivers with their heads at low altitude will slow and their banks would rise accordingly, affecting local populations. Most rivers, however, are fed from higher ground and will likely not be affected to any great extent. As an example, Lake Victoria which, along with other smaller lakes, feeds the Nile River in Africa is at an altitude of 1135 meters, so we can expect little effect from a sea level rise of 1 meter at its exit. The Mississippi is fed by Lake Itaska at 450 meters. The Amazon and Danube Rivers are fed from sources high in mountains.

Second order effects may also be damaging. Infiltration of salt water into areas that normally contain fresh water is inevitable, affecting wildlife habitats, drinking water, and coastal infrastructure.

EXTREME WEATHER AS IT AFFECTS HUMANS

In this section I will start by briefly summarizing the main conclusions from the ICPP report "Long-term Climate Change: Projections, Commitments and Irreversibility". (Collins, 2013).

In Chapter 2 it was shown that the expected increase in precipitation had not yet materialized, except possibly in the Northeastern United States. As temperatures increase into the future, however, increasingly warmer ocean and air temperatures must inevitably produce a more active water cycle. Below are charts showing projected precipitation increase.

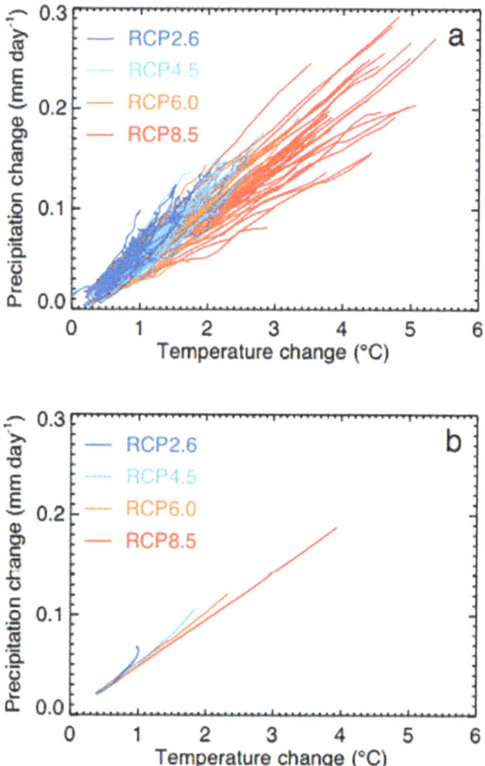

The first chart shows the results from different climate models and includes statistical variance. The second one shows the mean lines for the 4 RCP scenarios. Note that these charts show changes in mm/day per each 1°C change in global average temperature. Below is shown percentage change as compared to the base period of study, 1986-2005.

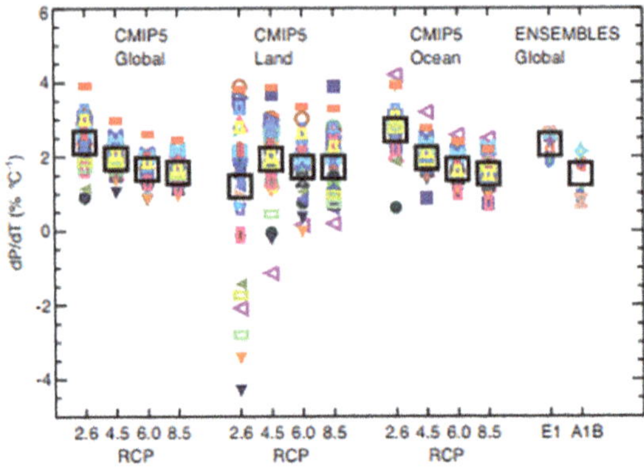

The conclusion is that around 2% increase in precipitation can be expected for each 1°C increase in temperature. This does not seem too dramatic, but below are images that give more interesting projected regional effects, for the most extreme RCP8.5 scenario. The acronyms DJF, MAM, JJA, and SON refer to the seasons (example DJF = December, January, February).

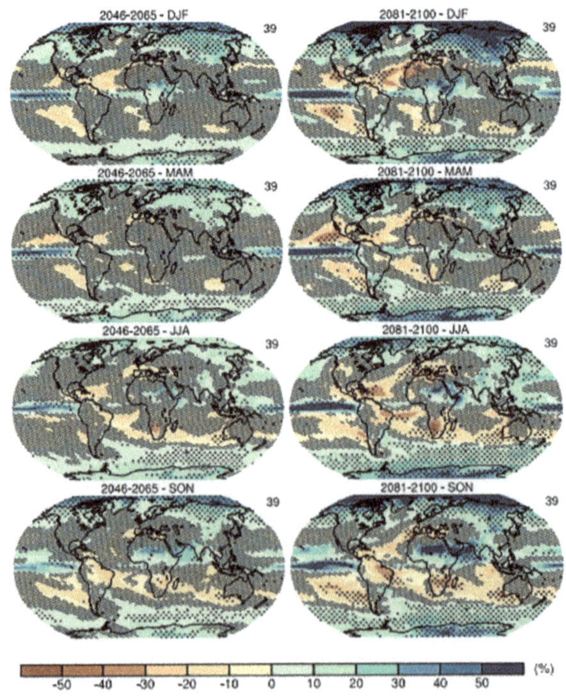

Projections are considerably more dire when regional differences are considered. We see areas in the Southern hemisphere where rainfall is down by as much as 30%, and in the far north they are up locally by more than 40%. This would indeed be a different world. We must keep in mind, however, that there is considerable uncertainty associated with these predictions.

The large increase of snowfall in the Arctic winter is of course a positive outcome, because it can attenuate the loss of ice in summer months. But overall, the models still show net loss of water from the Arctic, as discussed in the section on sea level rise.

In a similar way, temperature increase of up to 4°C may not seem too dire, and in some areas perhaps it would be welcome. However, that is a global average and does not include temporary spikes. Below are projections for peaks and valleys of the temperature scale, with regional differences also included, for the RCP8.5 scenario on the left. On the right are the global averages for the 4 RCP scenarios.

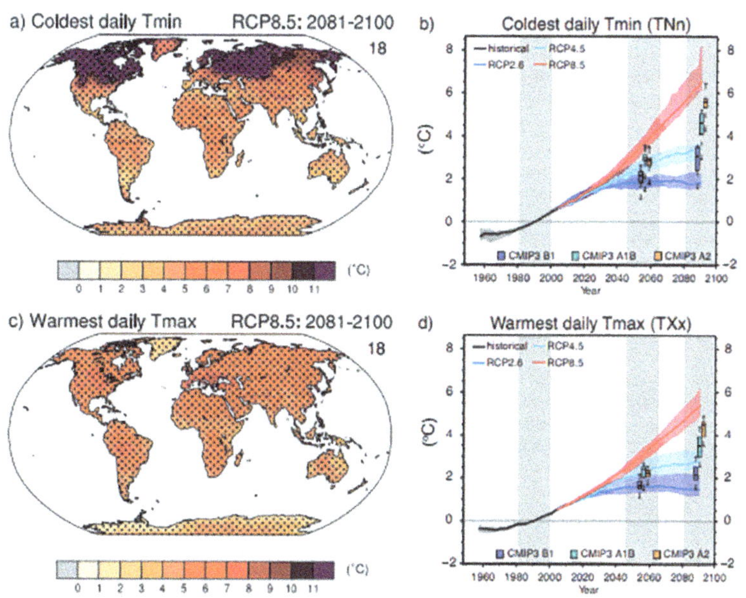

People in Canada and Russia might be happy to see that those worst cold days in winter are expected to be up to 10°C warmer. No-one in the world is likely to be pleased, however, with temperatures on peak days climbing 7°C or 8°C.

Next, I was surprised to find that an attempt was made to predict future cloud cover, which intuitively I thought would be quite impossible. Blue-Grey indicates increased cloud cover and Yellow Orange indicates reductions.

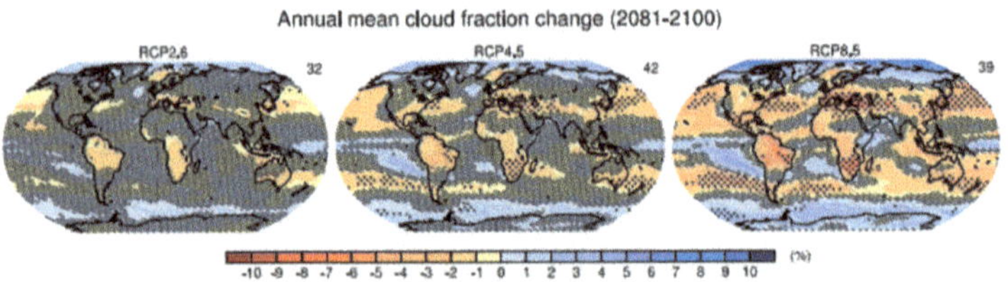

Cloud cover is critically important because clouds are a net cooler of the planet. They block out some of the sun's radiation, which in the total heat balance outweighs their trapping of heat radiated from the Earth. The net radiative forcing of clouds in the current world is around -20 W/m^2.

In the RCP8.5 scenario we see significant regions where cloud cover would be reduced, leading to further heat input to the planet. It should be pointed out that uncertainty on these predictions is quite high. Cross-hatched areas have mean changes that are greater than one standard deviation in the uncertainty band.

Regarding relative humidity (RH), the conventional wisdom is that it should not change greatly with increasing temperature. This is because the air will naturally absorb more water as it heats. The projections go further, however, to include second order effects of changing cloud cover and atmospheric convection.

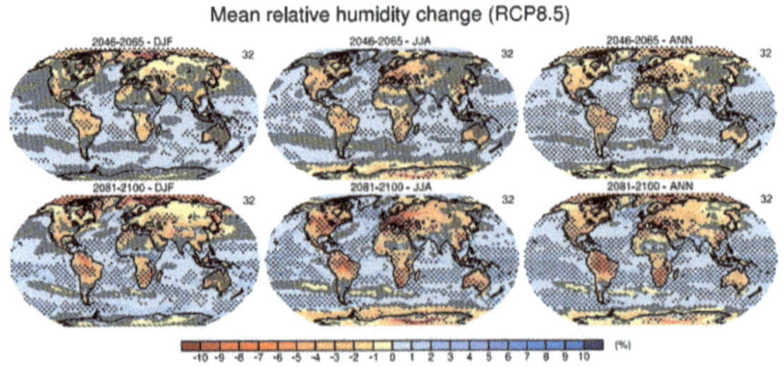

We see that over the ocean, where most of the water vapor source is located, RH goes up, but over land it goes down. This would certainly aggravate droughts in regions that are already dry, as the air will more readily draw moisture from the Earth.

There are no detailed projections for severe storm activities (cyclones, hurricanes, typhoons). There is, however, a qualitative assessment that increases in such activity are not likely, and in fact there is a small probability of a decrease in such storms. There is also a tacit agreement that in current times the intensity and frequency of storms are likely not greatly changed from historical behavior, a conclusion that I had come to independently in Chapter 2. This is astonishing, because in the media almost every extreme storm event is causally linked to anthropogenic Climate Change. This gives me more confidence that the IPCC has remained technically sound, allowing factual observations and robust analysis to draw conclusions. Similar conclusions are drawn for floods and droughts in the current world.

In the future world, drought is expected to be aggravated in dry areas, negatively affecting agriculture, and drinking water security, mainly in lower latitudes. In higher latitudes, increased precipitation is expected to cause flooding along riverbanks.

EXTREME WEATHER AS IT AFFECTS ANIMALS AND PLANTS

The IPPC reports predict high levels of stress onto many animal species, especially in tropical regions. Historical data on animal extinctions show that they have occurred even with environmental changes that were much slower to develop than rates predicted for future climate change. Many animal species will naturally migrate to find cooler climates. In principle plant species can also "migrate", but at speeds much too slow to outrun the effects of climate change. Warming in the Arctic, which is projected to be greatly amplified as compared to the rest of the world, will likely cause stress onto some species, but some may adapt to higher temperatures.

Stresses caused by climate change are in addition to the other stresses caused by human activity, which are already causing damage at rates so high that we are considered right now to be in the middle of the 6th mass extinction in Earth history. The prime culprits in today's world are habitat destruction, hunting of endangered species, and overfishing.

With the exception of a follow-up from Chapter 2 on insect and ectotherm populations, I will not delve here into effects on individual land-based animal species because they are difficult to predict with reasonable accuracy. I would, however, like to go into some detail about the effects of ocean acidification on calcifiers. Calcifiers are marine animals that build shells or exoskeletons. Examples are plankton, shellfish, and coral reef builders. A brief description of the science is in order.

The oceans absorb a great deal of the CO_2 in the air, and this has been placing limits on global warming. CO_2 reacts chemically with water, however, to form carbonic acid. The acid in turn releases hydrogen ions into the ocean, making it more acidic. Acidity is measured using the PH scale to determine the concentration of hydrogen ions. The hydrogen ions bind to carbonate (CO_3^{2-}) to form bicarbonate (HCO_3^{-}), which decreases the amount of carbonate available for calcifiers to form their shells.

Water is neutral on the pH scale with a value of 7. The ocean is slightly alkaline (opposite to acidic) with a pH of 8.06. The pH scale is logarithmic, so any change of 1 point implies a change in acidity or alkalinity of ten times.

Many calcifiers such as plankton are at the base of the ocean food chain. Fish and mammals in the ocean therefore rely on them for their survival.

The image below (Li-Qing Jiang, 2019) shows the expected progression of ocean PH level in the most severe RCP8,5 scenario.

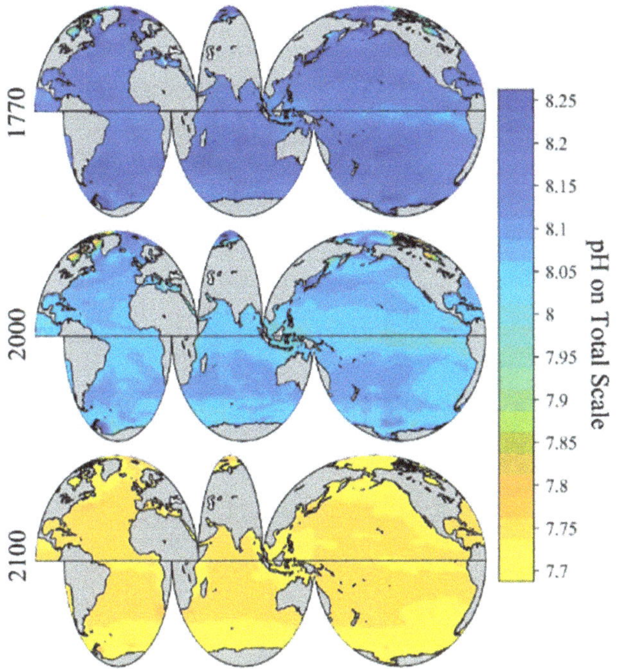

The change in pH from 8 to 7.7 means an increase in acidity of around 100%.

There is great difficulty in assessing the impact of acidification. In natural environments, excess acidification has not yet occurred, except in some small regions where thermal vents from the ocean floor, or volcanic activity, is locally releasing excess amounts of CO_2 into the water. Some scientists have attempted to artificially introduce increased acidity into coral reefs areas. There are also some laboratory-level experiments reported where input and output parameters can be well-controlled.

In this laboratory experiment (K. R. N. Anthony, 2008) A group of tanks containing 3 species of coral were filled with flowing unfiltered sea water, and then subjected to varying levels of temperature and CO_2 infiltration, over an eight-week period. Additionally, light was provided over simulated daytime conditions and removed at night.

Three things were measured: the extent of bleaching, the productivity of reproduction of the coral's algae and calcification (measured as weight gain).

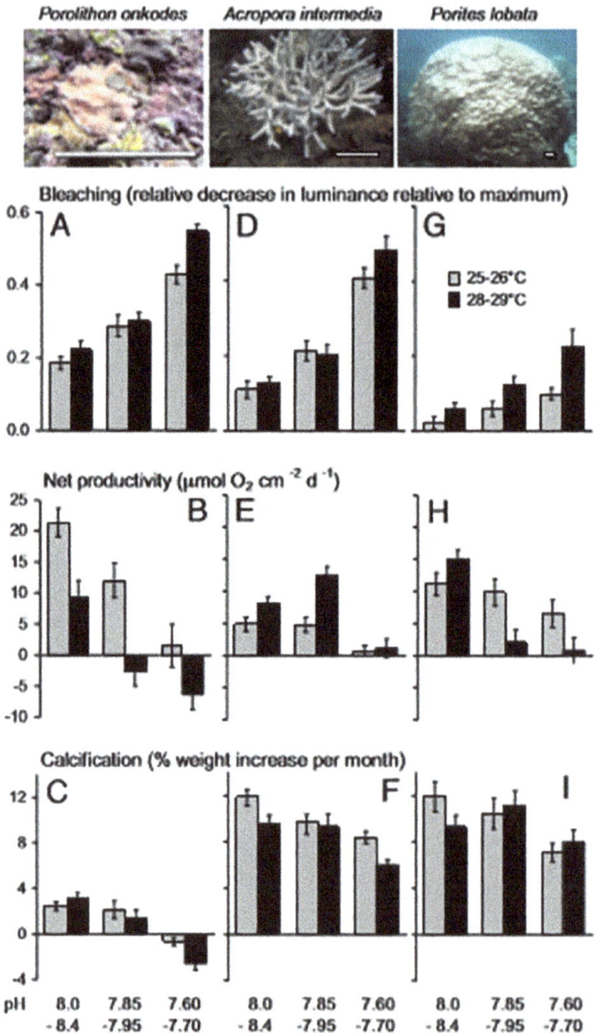

The baseline conditions, represented by the left–most bars on each graph, are those that were prevalent at the time of writing: a sea surface temperature range of 25.5°C to 28.5°C (including, I think, winter to summer variations), and CO_2 infiltration into water representative of the atmospheric concentration of 380 ppm, and a pH of 8.2. For CO_2

content, 2 other future conditions show pH levels of 7.9 and 7.65, the latter value being close to the projected pH in year 2100 with RCP8.5.

It can be seen that the addition of 3°C had little effect on bleaching. Perhaps the peak temperature had not reached the presumed tipping point. For algae productivity, the addition of heat was strongly negative in one species and moderately positive in the other two. The addition of acidity had strong negative effects on bleaching and productivity, and moderate negative effect on calcification.

In one IPPC report (Field, 2014) the following table is produced, summarizing many such studies and their outcomes.

Taxon	No. of studies	No. of parameters studied	Total no. of species studied	pCO_2 where the most vulnerable species is negatively affected or investigated pCO_2 range[a] (µatm)	Assessment of tolerance to RCP 6.0 (confidence)	Assessment of tolerance to RCP 8.5 (confidence)
Cyanobacteria	17	5	9+	180–1250[a]	Beneficial (low)	Beneficial (low)
Coccolithophores	35	6	7+	740	Tolerant (low)	Vulnerable (medium)
Diatoms	22	5	28+	150–1500[b]	Tolerant (low)	Tolerant (low)
Dinoflagellates	12	4	11+	150–1500[b]	Beneficial (low)	Tolerant (low)
Foraminifers	11	4	22	588	Vulnerable (low)	Vulnerable (medium)
Seagrasses	6	6	5	300–21000[a]	Beneficial (medium)	Beneficial (low)
Macroalgae (non-calcifying)	21	5	21+	280–2081[a]	Beneficial (medium)	Beneficial (low)
Macroalgae (calcifying)	38	10	36+	365	Vulnerable (medium)	Vulnerable (high)
Warm-water corals	45	13	31	467	Vulnerable (medium)	Vulnerable (high)
Cold-water corals	10	13	6	445	Vulnerable (low)	Vulnerable (medium)
Annelids	10	6	17+	1200	Tolerant (medium)	Tolerant (medium)
Echinoderms	54	14	35	510	Vulnerable (medium)	Vulnerable (high)
Mollusks (benthic)	72	20	38+	508	Vulnerable (medium)	Vulnerable (high)
Mollusks (pelagic)	7	8	8	550	Vulnerable (low)	Vulnerable (medium)
Mollusks (cephalopods)	10	8	5	2200 (850 for trace elements)	Tolerant (medium)	Tolerant (medium)
Bryozoans	7	3	8+	549	Tolerant (low)	Vulnerable (low)
Crustaceans	47	27	44+	700	Tolerant (medium)	Tolerant (low)
Fish[b]	51	16	40	700	Vulnerable (low)	Vulnerable (low)

Many species actually show beneficial effects of increased acidity, while many are shown to be vulnerable.

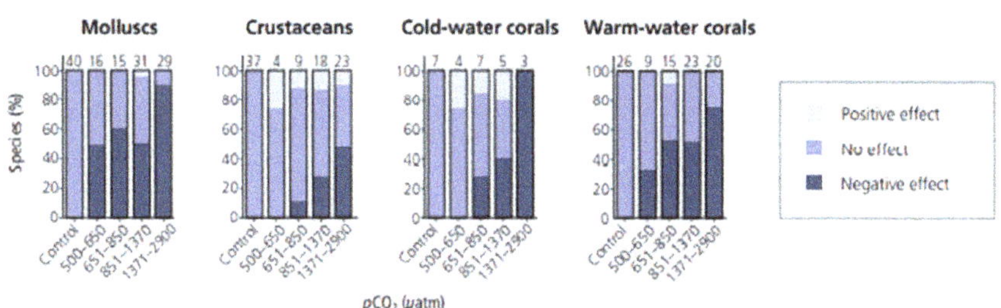

Evidently there is a great deal of variability, but all species show negative effects at higher acidities.

There are other studies available that have found tolerance to higher levels of acidity among corals when multiple generations of exposure are allowed to occur. Indeed, since we are looking at many decades of gradual build-up of acidity, there is always the possibility that positive adaptations could take place.

While small calcifiers form the foundation of some food chains in the ocean, insects are the base for many land-based animals. Recall the following graph from Chapter 2, where fitness is defined as fraction of growth rate by reproduction.

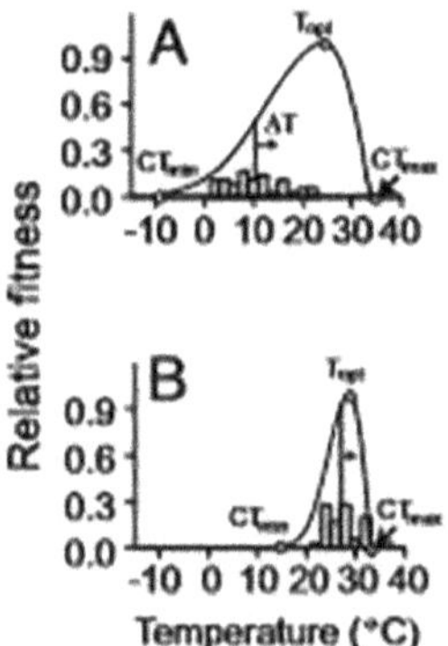

'A' represents tolerance to temperature variations for insects in Northern latitudes, and 'B' is the equivalent for insects in the tropics. The conclusion is that insects already adapted to large temperature swings in the North will adapt more effectively to climate change, while those in the tropics that are adapted to a small temperature range will suffer more.

Below, the study included projections to year 2100 for both insects (A) and some of the animals that eat them (C), as well as a global map of fitness fraction.

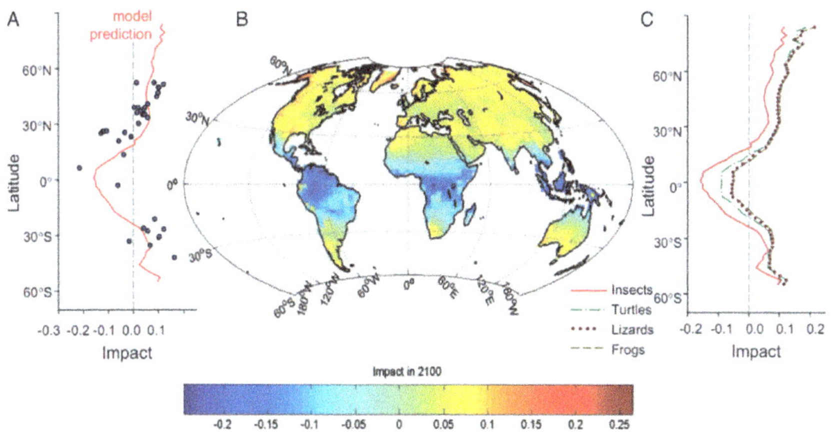

Many species see positive impact of climate change and might be expected to be more numerous in high latitudes, while those in the tropics, where more diversity exists, will see declines in populations.

CLIMATE CHANGE IN THE PUBLIC EYE (PART 2)

As discussed in Chapter 2, the media blitz around climate change has entrenched into the public consciousness that strong negative effects of climate change have already arrived, which is demonstrably not true. This environment has perhaps unintentionally attenuated fears over projections of future dire consequences which, based on evidence summarized in this chapter, will most likely come true. Perhaps people feel that they are already living with it, and so far, their lives are not affected, so why worry?

For predictions of the future, articles in the media tend to be anchored in the very studies that the IPCC has used, or will use, to set projections of the effects of future climate change. These are mostly scholarly papers that are firmly grounded in science. For that reason, the media are somehow often getting it right. There are still many on-line news services that relish in the latest dire news and tend to inflame the debate. On the other hand, with time, climate change denial is withering away.

CHAPTER CONCLUSIONS

If we assume the worst scenario with little mitigation of emissions, we see a world in 2100 that is significantly warmer on average, and extremely hot on worst (peak) days. Regions of the Earth that currently have extremely hot climates may reach beyond limits that are bearable by humans. Precipitation will increase greatly in many areas and reduce in some. Arid regions will get dryer due to high temperatures and reduced humidity. The Arctic will get much warmer than other regions, and there could be a complete loss of summer sea ice. Sea levels could rise by around 1.5 meters, necessitating the displacement of many millions of people, and a complete re-imagining of coastal infrastructure. It is almost certain that food production will be affected in many regions due to drier soil and intolerance to heat. Increased flooding along certain riverbanks can be expected. However, more severe Hurricanes and Typhoons are not anticipated.

Many animal populations already affected by other human-driven stresses will see more difficulty. Migrations to cooler climates are probable, and so are extinctions. Animals in the tropics, especially corals, insects, and insect predators, will see large declines in population. Insects in higher latitudes may see increases in population.

Regardless of mitigation scenarios, it is highly likely that many of the negative aspects of climate change may continue to worsen into the 22nd century.

Also worrisome is the fact that climate models, in spite of being very sophisticated, still have broad confidence bands. This means that things could be even worse. Conversely, they could also be better.

The strongest influence of all will be the extent to which mitigation plans can be implemented.

FOSSIL FUELS CONSUMPTION AND SUPPLY

IN THIS CHAPTER WE WILL look at some details about fossil fuels which need to be understood before going forward with any plans for mitigation of climate change. I remind the reader that in Chapter 1 we saw that fossil fuels are responsible for only 65% of damaging greenhouse gas effects, so this chapter deals only with that portion.

POPULATION AND URBANIZATION

At the beginning of the 19th century, in pre-industrial times, the world had a population of about 1 billion.

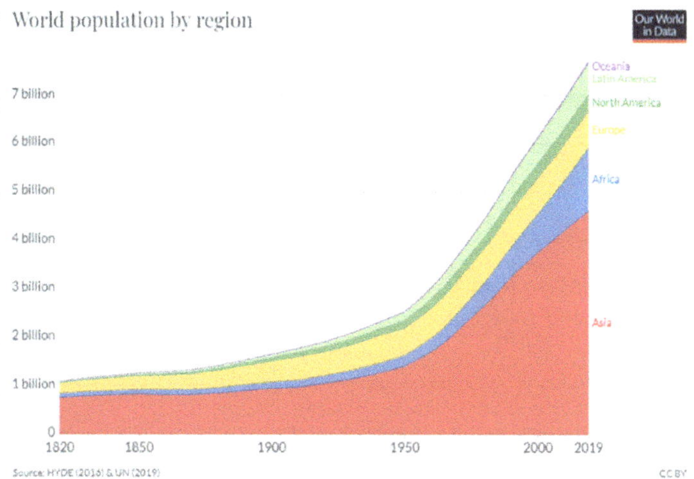

More than 95% of people lived in rural areas, and more than 50% were employed in agriculture. Today, in the heavily industrialized countries in Europe and North America, that number is a paltry 2%. However, it is still extremely high in India and sub-Saharan Africa.

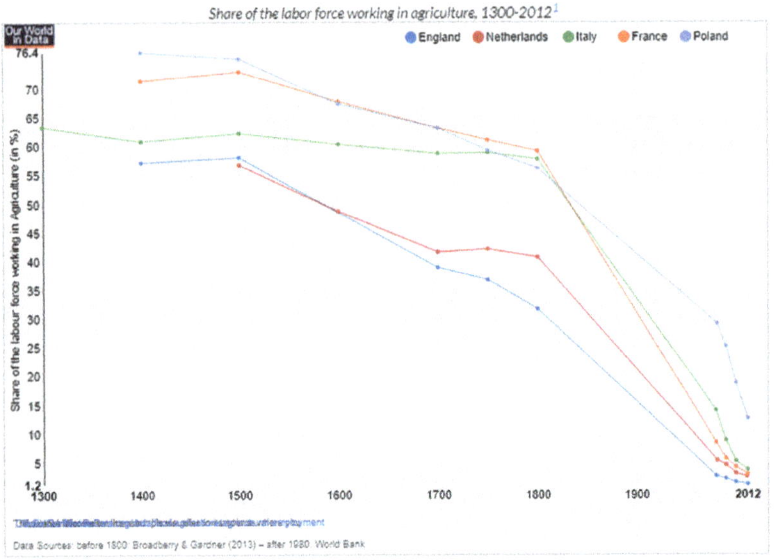

Meanwhile, urbanization has progressed to a level exceeding 50% world-wide, and to as much as 80% in North America and Japan.

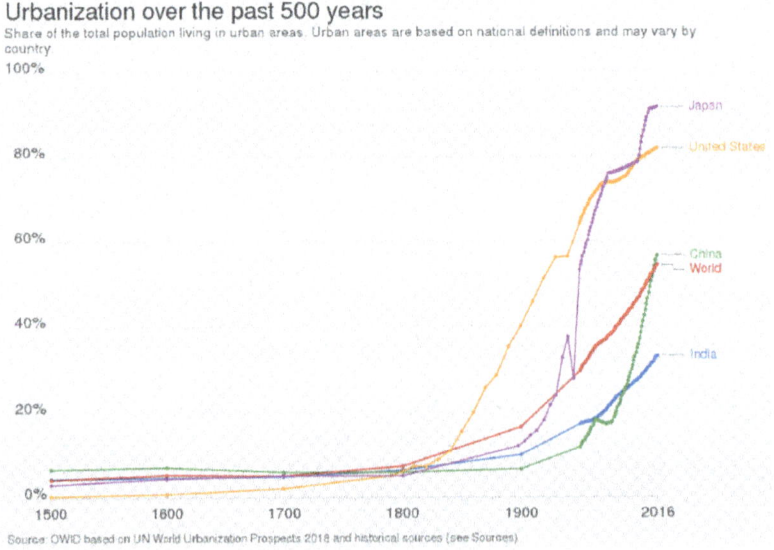

Regional population growth follows a pattern similar to urbanization, but with some important differences.

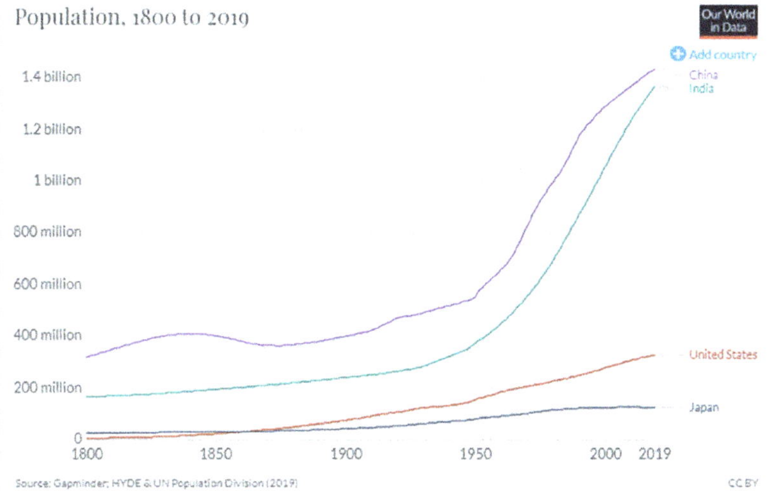

India has reached only 30% urbanization while having a population explosion, indicating that its labor is still tied heavily to agriculture. USA's population has progressed at a much lower rate, but the transition to urban areas has been enormous. This indicates extreme industrialization, including the mechanization of agriculture.

Looking world-wide at growth rates, we see below that all regions are trending downwards except for Africa.

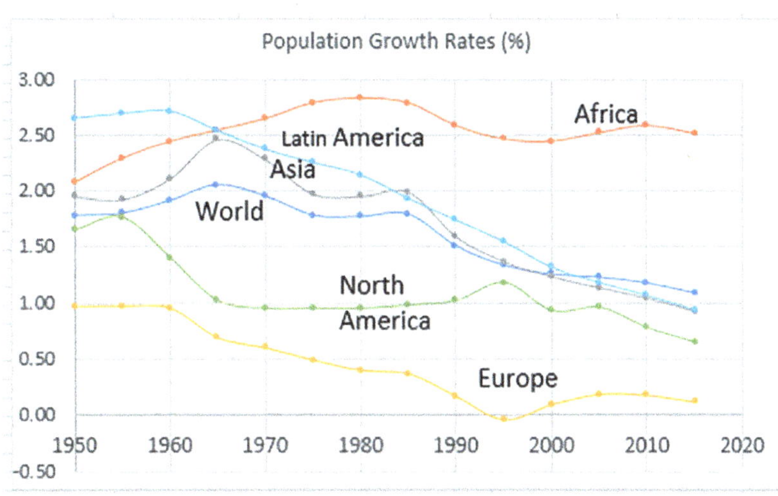

The first industrial revolution was already ongoing in 1800, to the extent that coal was used to run steam engines and to fire ovens for forging iron. From 1870 the second industrial revolution started, which gradually introduced steam powered transportation, the use of oil supplementing or replacing coal, electrification, and the internal combustion engine. By the end of WWII, much of the world was ready for a complete transformation of society. The charts above show that after 1950 the trend towards urbanization was greatly accelerated, fueled by reduced need for agricultural labor in the countryside, and increased need of industrial labor in the cities.

A question often asked is: did population explosion drive the heavy increase in demand for fossil fuels, or did the availability of fuels and their energy-extracting machines drive the population explosion? The answer is probably yes to both, but the I think the 2nd question is the most important. Amid all the complexity, there is one simple axiom: the world could not feed 7 billion people in pre-industrial times. The limits of agriculture using human and animal labor were reached at a population of about 1 billion people. The mechanization of farming, and the subsequent migration to the cities, I contend, were the results of motive power using fossil fuels.

What does this mean for us today? It means we cannot go back. The quaint notion that we can revert to "a simpler life" with organic farming in our own backyards is absurd. Firstly, most people do not have back yards, and if they did, they would see their meagre crops stolen by neighboring hungry families. Factory farming has taken over, and the distances that food must travel to reach hungry city dwellers cannot be traversed with anything but fuel-powered vehicles. Furthermore, the productivity of these farms is enormous, producing much more food per unit area than any time in history. Much of this productivity is unfortunately driven by heavy use of fertilizers, which are disrupting Nature's nitrogen cycle. These fertilizers also require fossil fuels in their production.

In the 1970's, dire predictions were made of a massive food shortage if the world's population continued to creep above the 3 billion level. At the time there seemed to be no technical solution. Such is the improvement in agricultural productivity that today there is a surplus of food for 7 billion people. Obviously, it is badly distributed, with residents of the wealthiest countries wasting much of their food. In principle, however, the world's farms are capable of feeding everyone, if we can only redistribute that capability more fairly.

It can be argued that in some developing countries there is still a large percentage of the world's population using older methods to grow food. But this still relies on heavy manual labor and essentially low productivity of the people involved. Is this sustainable? Probably not.

In Industries other than agriculture, motive power also brought tremendous improvements in productivity. Imagine the plight of a construction worker prior to the existence of fuel-powered machinery? How much can a person achieve in a day's work if he has only his or her 2 hands to move and build? Massive numbers of people were needed to build things, and/or the building of large structures took decades to accomplish. Compensation of these workers was minimal, and not only because owners and governments exploited them. The fact is that worker productivity was inevitably extremely low. Today, in any of the heavily industrialized countries, compensation for a skilled construction worker is sufficient to support a family in relative luxury as compared to previous times, and included would be health care, sick leave, and greatly improved workplace safety.

Taking this argument to an extreme, consider that from ancient times until the Industrial Revolutions, slave labor was paramount in achieving any large public or private economic venture. It freed up slave owners from the difficult physical work, and compensation was limited to the bare necessities of sustaining life. The common narrative is that slave labor was gradually eliminated because of our moral and ethical enlightenment. But without the productivity afforded by fossil fuels, I wonder if it might have still survived to this day.

While we may regret the damage done to our planet by the burning of fossil fuels, perhaps we should also consider the benefits they have brought to mankind, benefits that most of us still enjoy and are evidently unable to give up. These benefits are not only those of convenience, like the ability to drive a car to go to the mall. They are also fundamental: food security, electricity supply, and the wealth needed to provide health care and safety to most of the world.

FOSSIL FUEL CONSUMPTION

In this section I will gratefully make liberal use of data from a very excellent book by Simon Pirani called "Burning Up", published in 2018. This book describes in great detail the history of fossil fuel entrenchment into our societies, with both voluminous raw technical data and commentary on the social implications in "rich" and "poor" countries. (Pirani, 2018)

In concert with the population and urbanization graphs from above, here is a chart describing the rise of fossil fuel consumption during the Industrial Revolutions. For the sake of illustration all types of fuel have been converted to Coal-Equivalent. Units are Millions of Tons.

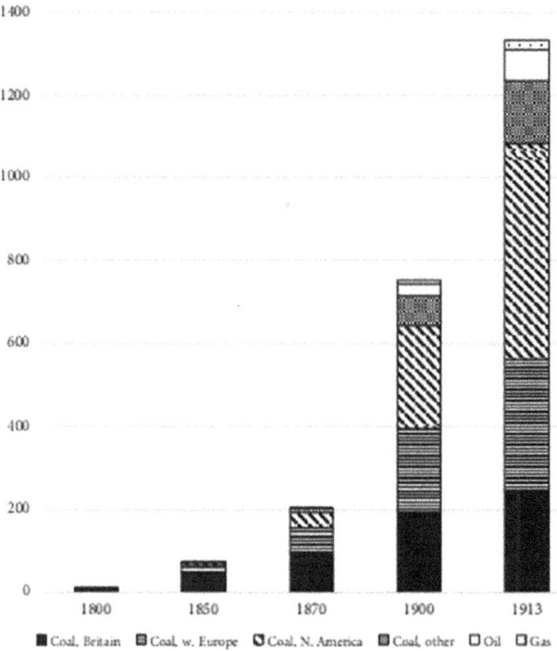

Note that the category "Coal, other" represents the world outside of Europe and North America. It shows that the rest of the world experienced a lag in the incorporation of fossil fuels into their economies.

The initial stages of this revolution were probably not very pleasant for city-dwellers. The air would have been badly polluted by burning coal. Workers initially received low wages and employment was very much at the whims of the factory owners. Workplace safety precautions were dismal. Somehow this all still seemed preferable to subsistence living in the countryside, and exodus to the cities continued.

Over time, the tremendous increase in wealth brought on by Industry filtered its way down, in part, to factory workers. Furthermore, the newfound wealth and population density created new blossoming businesses in the Services sectors of the economies.

Gradually, the internal combustion engine became the preferred tool for transportation, but coal remained as a staple for industry and electricity generation. By 1950, fossil fuel consumption had doubled again as compared to the amount shown above in 1913, and the USA was already the dominant user. In most of Asia and Africa, the "lag" in usage of fossil fuels continues to this day, where there is still prominent reliance on biomass (mostly wood) for cooking and unmechanized farming.

Below are charts showing progress of population and fossil fuel consumption in 3 countries since the start of the era of global warming (1980). Population scale on the right is in millions (solid lines), fuel scale on the left is in megatonnes of oil equivalent (dotted lines).

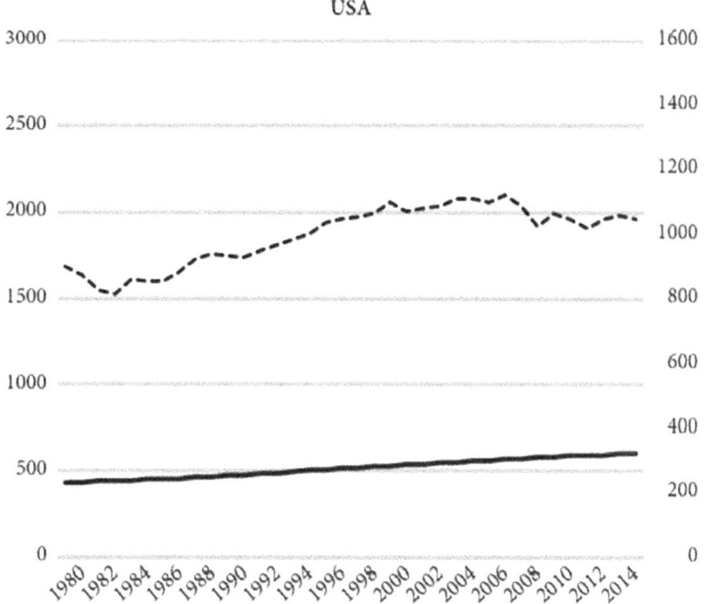

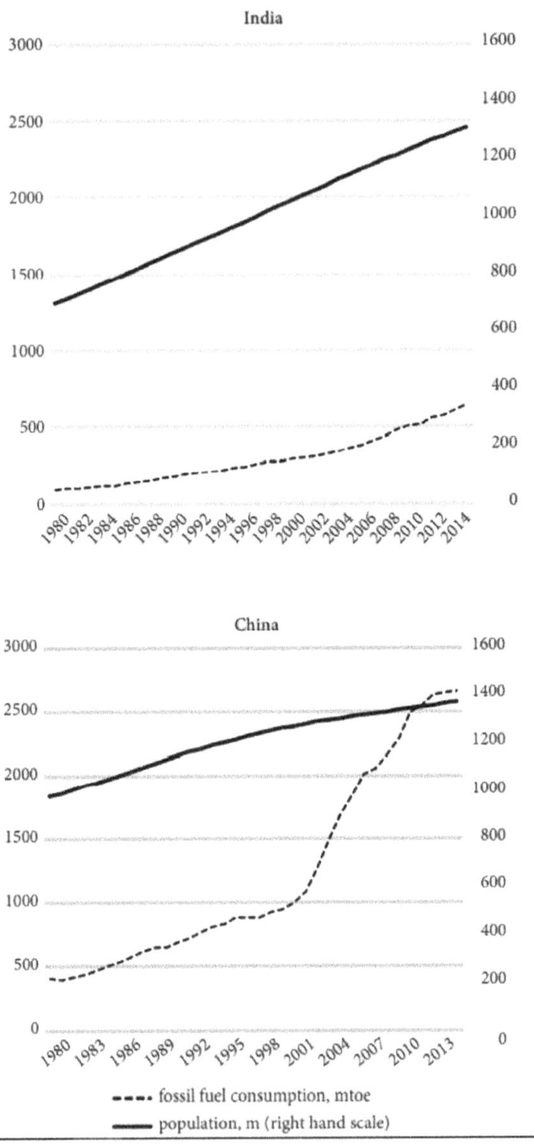

India is ramping up very slowly, while China is increasing consumption at a high rate, exceeding USA Consumption in total, but still behind on a per capita basis. Interestingly, USA consumption has leveled off, possibly a small bright spot for the future.

It seems that fuel consumption and prosperity go hand in hand. While India's progress in bringing even basic needs to all of its people is slow, China has seen an increase in average income following its meteoric rise in consumption. Since 1978 China's Gross Domestic Product (GDP) has seen a 10% year-over-year increase. The World Bank now considers China to be a medium income country, and 850 million people have been lifted out of poverty. The message here seems to be that the increased productivity associated with fossil fuel-driven industry brings wealth to the people, not only in the era of initial industrialization but also in today's world. If this is true, it does not bode well for the future. It seems, so far, that avoiding serious climate change and improving the prosperity of impoverished nations might be mutually exclusive. It will take powerful forces to change this trend.

We saw in Chapter 1 that on aggregate global CO_2 emissions have been increasing at greater rates than population. The charts above show that for individual countries there is a great deal of variability, depending on the stage of development. Intuitively, a better indicator might be GDP.

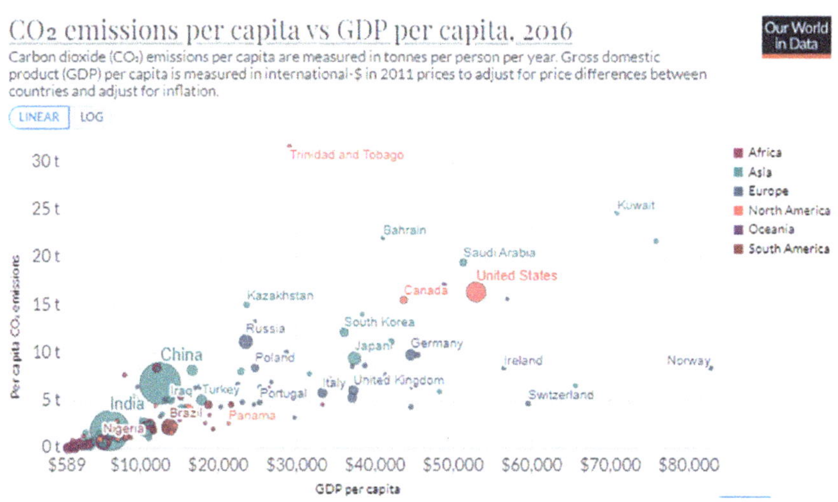

Indeed, there is a trend towards greater consumption with greater economic activity, but there is also a big scatter at any GDP level, indicating that some countries are using energy much less efficiently. USA and Canada stand out as energy hogs, while wealthy European countries are managing their resources much more efficiently.

In the Pirani book there is an astonishing anecdote that I will relate here. The cities of Atlanta and Barcelona have similar populations and GDP, yet Atlanta uses 11 times as much energy for transport. Atlanta has a maximum distance between two points of 137 kms, while in Barcelona it is 37 kms. Atlanta has a downtown core dedicated to business. People live primarily on the outskirts in suburbs that over time have sprawled away from the core. Cars are used for virtually every trip, be it travel to work, shopping, or entertainment. In Barcelona, like many European cities, people live, work and play in their neighborhoods, and there is an extensive public transit system. Most large US cities follow the Atlanta model, and to some extent Canada's larger cities do as well.

What should be done? Can we make our cities less energy-intensive? Can we find ways to avoid developing countries from becoming energy hogs while still developing their economies? China is showing a huge ramp-up in consumption tied to prosperity. What if the same were to happen in India? It would dwarf the consumptions of all other nations. It is likely we can find only partial solutions.

Below is the total history of global consumption of fossil fuels.

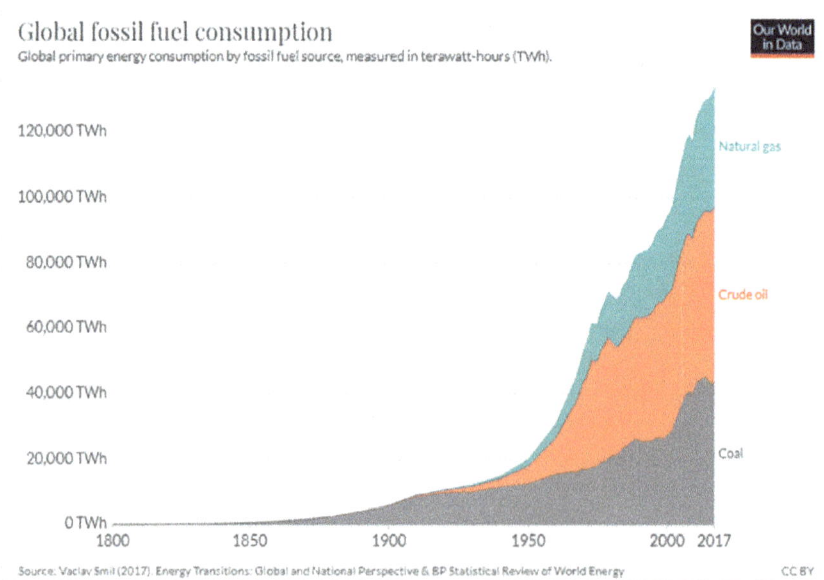

Today coal is used primarily for electricity production, crude oil for motive power, and natural gas for heating and cooking, but there is considerable overlap in these functions.

The ramp-up is dramatic and does not show signs of attenuating.

THE SUPPLY SIDE

With ever increasing consumption, is it possible that the supply can run out before the worst effects of global warming are even realized? The IPCC considers the concept of a carbon budget. In order to limit warming to any temperature level as compared to preindustrial times, there is an associated amount of CO_2 that can be emitted. Then, knowing the amount of CO_2 emitted for a given amount of fuel consumption, we can determine how much of the world's remaining fuel reserves is required to reach any target temperature level. I have used IPPC data for CO_2 emissions as follows: Oil .42 tons/barrel; gas 59.7 tons/million cubic feet; coal 1.81 tonnes/tonne (bituminous) and 1.20 tonnes/tonne lignite, with a bituminous/lignite assumption of 70%/30%. For fuel reserves, I have used data from British Petroleum (BP) Statistical Review of World Energy (2018).

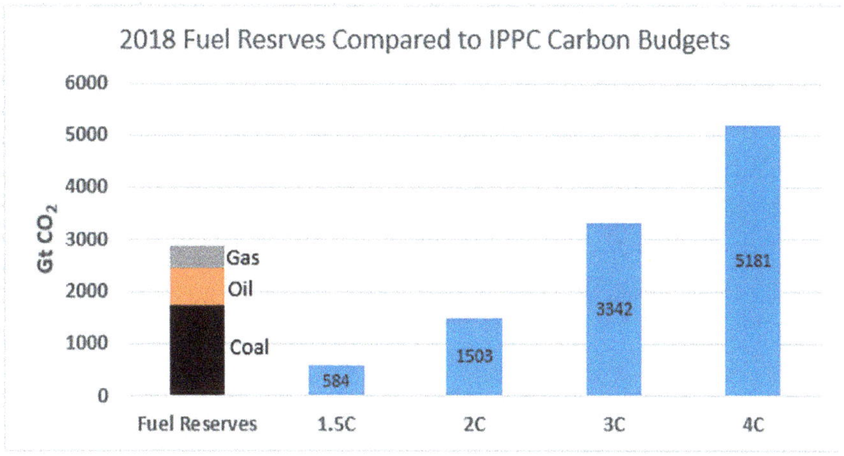

There are two important assumptions built in here: consumption rate continues as it is today, and no new reserves are found. From a consumption point of view, there has been a continuous increase in rate year over year, so it could be argued that reserves might deplete sooner than shown here. On the other hand, we are meant to greatly reduce consumption to achieve climate goals, therefore reserves in that scenario would go on for longer than shown.

With warnings that any increase greater than 2°C would be catastrophic, some analysts have shown similar charts and concluded that we should stop looking for new reserves, and even leave some existing reserves in the ground. In this way we are forced to never exceed 2°C of warming. In all cases, however, no plan is offered on how we replace these fuels with other sources of energy, or how we reduce our consumption.

In Chapter 3 we reviewed the future world mostly for RCP8.5, which is the worst scenario where little mitigation of emissions would occur. In that scenario, the temperature increase by year 2100 was 4°C. The chart above shows, however, that we will run out of fossil fuels before such a scenario could ever occur. In fact, if we assume current rate of consumption, the estimated time to depletion of reserves is: Oil and gas, 50 years; coal, 132 years. Coal has a large reserve, but it cannot easily be used to replace oil and gas in many applications, and its rate of CO_2 emissions per unit of energy produced is much higher.

This means that global warming is not the only driver to reduce fossil fuel consumption, and it is possibly not the most serious one either. Unless we find ways to replace the functionality of these fuels, we are in for a painful transformation of our societies if we run out.

It is evident that the supply of fossil fuels is not unlimited. Already we are extracting deposits through processes that are more difficult and expensive than before. The Canadian Tar Sands were thought at one time to be impractical to use and too expensive to consider when oil prices were low. With the increases in price, it became possible. High pressure steam is used to bring bitumen to the surface, and then more energy is spent to refine the bitumen into usable oil. It is estimated that between 20% and 30% of the usable oil's energy is spent to extract and refine it, compared to around 4% with a conventional oil well. In the United States Natural Gas and Oil are being extracted from shale deposits using fracking, a process whereby pressurized water containing sand is used to shatter the rock, releasing trapped hydrocarbons. This has made the USA the World's leading producer, although it is considerably more expensive than tapping an underground chamber already filled with oil. In the Arctic it is known that there is a large potential for oil and gas exploration, but mostly this is not exploited for environmental reasons.

I will explore the subjects of currently available alternative energies in Chapter 5. We will see there if it is possible to ween ourselves off of fossil fuels in time to stay within the 2°C limit.

CONCLUSIONS FROM THIS CHAPTER

Fossil fuels and the growth of population and wealth have been inextricably intertwined. There may be no going back on the benefits because they include fundamental necessities of life, including food security for 7 billion people.

History has shown that fuel consumption and Gross Domestic Product have been causally linked. If developing nations follow the same path, as China has most recently done, then consumption will continue to climb unless alternatives are found.

Supply is limited. At current rate of consumption, oil and gas will run out in about 50 years. By that time temperature increase relative to pre-industrial times would be about 3°C. Unless alternatives are found, this could be a catastrophe greater in magnitude than the warming of the planet.

CHAPTER 5

MITIGATION EFFORTS IN TODAY'S WORLD

IN THIS CHAPTER WE WILL explore the efforts being made to implement practices that would reduce our emissions of greenhouse gases. The focus will be on alternative energy sources to reduce carbon dioxide emissions that are active at the time of writing. I will address other future opportunities in Chapter 7.

STATUS OF RENEWABLE ENERGY SOURCES

The latest BP Statistical Review of World Energy (Report, 2020) shows the following breakdown of energy, with all sources normalized to Exajoules (Quintillions (10^{18}) of Joules).

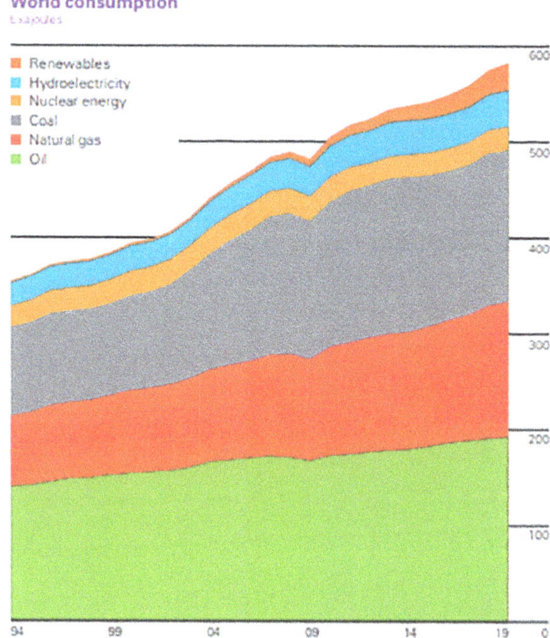

Renewables represents 5% of the total, although their rise over time as a percentage of the total is the highest. Renewables includes burning of biofuels and biomass, and these produce CO_2 in the same way as fossil fuels. The amount produced by wind and solar is slightly less than 4%. Therefore, we can say that the penetration of "clean" energy that leads to CO_2 reduction is about 4%. As a percentage of electricity production only, renewables represent 10%. Note that hydro power and nuclear power are not considered renewables in this accounting process.

This is an amazingly small number, considering that we have been talking about clean energy and renewables for at least 40 years. Clearly there are constraints that have limited their contribution, and we will explore those constraints in this chapter.

As I was writing this chapter, I was trying to do an accounting of the potential benefits of each of the alternative energy sources prevalent in today's world. I found it to be problematic because this requires projection into the future on what the demand will be. We cannot assume that today's demand for energy will stay constant over time, or even that it will continue to increase at rates consistent with historical values. Presumably, conservation efforts will start to have some impact on both the demand for energy and the relative proportions of fossil fuels and renewables.

For example, if we evaluate wind power and conclude that it can contribute up to 30% of electricity generation, we still need to know the total demand in order to determine what 30% means in terms of CO_2 reduction and temperature attenuation. Today it is 4% of a known number. Tomorrow it will be 30% of an unknown number.

Along the way I will be making some assumptions to fill this gap. I can get some help from the voluminous calculations done by the IPCC and its expert contributors. Recall the following chart from Chapter 3.

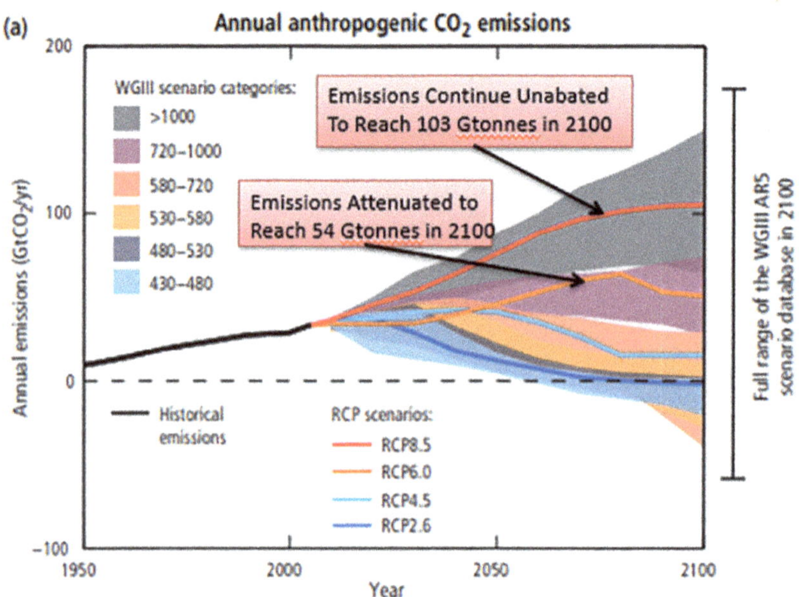

The RCP8.5 scenario assumes runaway emissions with no mitigation. It includes expert projections on energy demand. RCP4.5 and RCP2.6 assume drastic near-term reduction of emissions. All of these scenarios seem unlikely to me, so I choose where appropriate to follow RCP6.0. This scenario shows emissions levelling off in the near future and then rising only slightly over time. The energy demand is also built into this scenario, but emissions come down because it is expected that the Paris Accord and other future agreements will have this effect. The gap between the red and orange curves represents reductions in emissions caused by these commitments.

Next, I take the total energy demand from the BP chart above and project it into the future.

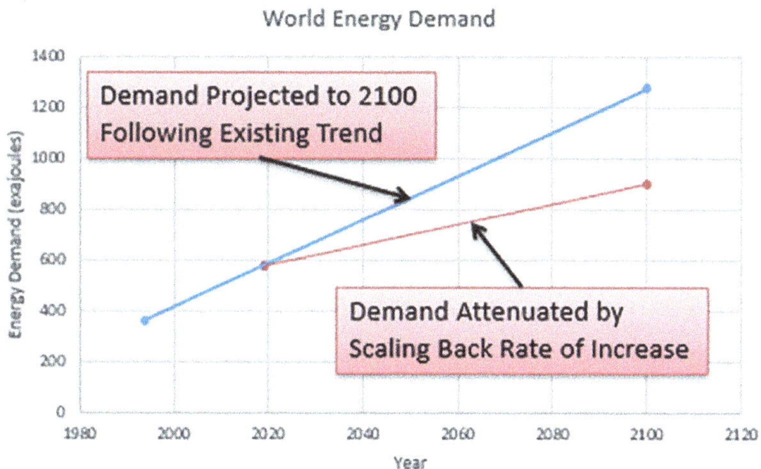

I have used the IPPC charts and made a best guess at reduced energy demand created by energy conservation activities, and also by the leveling off of predicted population growth. This is the line we will follow in trying to project savings in global temperature increase later in this chapter.

The constant rate of increase of demand is 3.95 exajoules per year. I choose this constant rate rather than a constant percentage, so the percentage increase goes from .68% in 2020 to .44% in 2100. This is a second level of attenuation due to levelling out of population.

I will assume that change in energy demand is similar in all economic sectors.

WIND POWER

Most of us have seen windmills popping up in the countryside, usually on farms and in unpopulated areas. The large diameter turbine blades are shaped to extract power from the wind, which generates rotation. This rotation in turn can be converted to electricity using large electromagnets. The prime locations are those with known high wind velocities, but no location is entirely reliable or predictable in this regard. Almost exclusively, wind farms are backed up by other conventional sources of energy for generating electricity. Otherwise, brownouts or blackouts would occur regularly, whenever the wind dies down.

Given that the obvious intention of wind power is to displace fossil fuel consumption, it must be done, it appears, without decommissioning conventional power plants, which will continue to be needed for low wind conditions. However, fuel inputs can be dialed down when wind power is available.

A few words on terminology are in order. Windmills are assigned a capacity factor, which is defined as the electrical output over time divided by the ideal output. The ideal output would occur when wind conditions are in a perfect state. Capacity factor must be measured at the specific installation location, and over a significant period of time like a full year. Even then, there can be year-to-year variations. Such are the vagaries of the wind. A typical capacity factor is 30%. This means that if the unit is rated at 2 MW, then we can expect it to deliver an average of .3 x 2 = .6 MW. It is useful to note that power extraction is roughly a square function of wind speed. In other words, half wind speed would deliver only one quarter of the power.

Most studies on wind power use this capacity factor in the calculation of power delivered to the grid, and then an associated savings in CO_2 emissions is calculated assuming that the fossil fuel is ramped down by the amount that would otherwise be used to generate the extra power. Many critics propose, however, that large fuel-burning plants cannot be ramped up and down without loss of efficiency. In other words, CO_2 emissions continue unless you shut the whole thing down. "Studies" can be found on anti-wind web-sites which claim that there is little or no net benefit of wind power, but I find that most of these are either lacking in supporting data, or use suspicious assumptions to arrive at negative conclusions. On the other hand, it is true that there is an important efficiency loss for power plants to operate with fluctuating supply, and mainstream supporters of wind power benefits for the most part ignore this factor.

As usual, I attempted to find some hard data to quantify this debit. I found a 2012 paper from the US National Renewable Energy Laboratory (NREL) (D. Lew, 2012) that looked specifically at all the effects of "VG" (Variable Generation) on fossil fuel-powered plants. Remarkably, there are real time measurements taken over the course of an entire year (2008) in the USA. In the following chart, CC, CT and GasStream are different types of natural gas-fired plants.

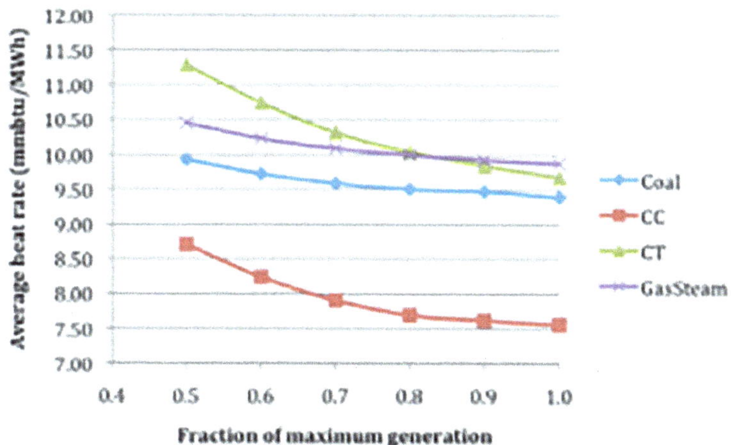

The heat rate on the vertical axis is directly convertible to CO_2 emissions. It can be seen that at part power there is a deterioration in efficiency that causes emissions to be higher than would be considered if we just use the capacity factor as an indicator. For example, at 50% power, which is a likely target reduction factor for fuel-powered plants, the emissions intensities increase by between 5% and 15%, depending on the fuel and the process. Still, considering that the other 50% of energy is being provided for free, this seems not to be a bad trade-off. It is especially interesting to see that coal emissions per MWh only increase by 5% when operating at 50% capacity. This contradicts many critics who claim that coal plants keep burning the same amount regardless of power extraction.

The curves on this chart are best fit to a considerable amount of scattered data. Here is an example of the raw data and the fit for CC units. These are natural gas-fueled gas turbine engines with exhaust heat recovery.

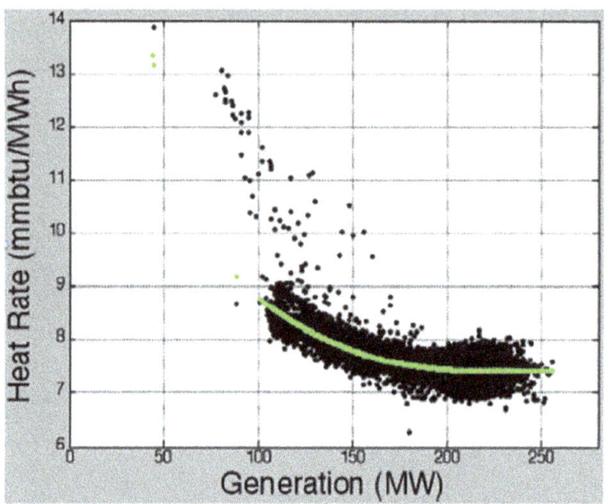

The scatter in the data reflects several variables, but one of them is likely the amount of renewable energy input to the system in parallel to this fuel-powered generation. In other words, over and above the deterioration in efficiency seen on the best fit green curve, there may well be another debit associated with variable input that is reflected in the data scatter.

Even without the existence of variable wind power, generating plants must cope all the time with variable output. Consider for example that the demand for electricity varies tremendously even over the course of a single day. Units must be ramped up and down, and some shut down and restarted, to meet requirements. Wind and solar certainly add to that burden, perhaps substantially. After all, the variations in demand are somewhat predictable using historical data, while the wind is not predictable at all.

In order to shore up the data, I located a European study (Miguel Angel Gonzalez-Salazara, 2018) which again uses real world data to determine efficiencies at part power caused by the introduction of renewables into the grid. "MCL" stands for Minimum Complaint Load, which is the load below which the facility should not operate due to restrictions like emissions.

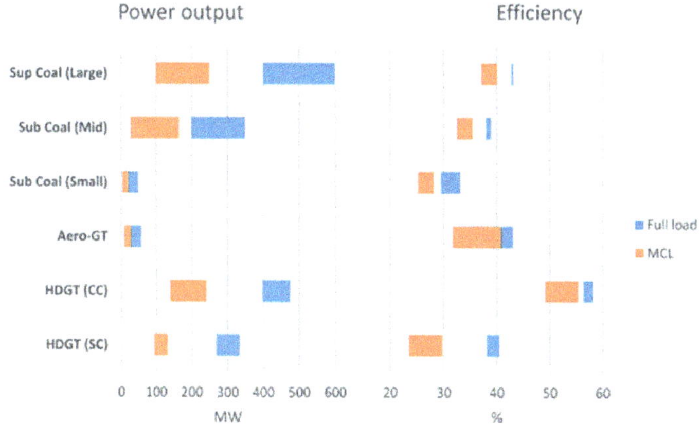

Similar to the Lew study, we see that coal has a smaller efficiency debit at part power than the gas turbine application, and the efficiency losses are similar, assuming that MCL is around 50% power. Here we have verification from which we can perhaps derive some conclusions for a calculation I will attempt later on.

Below are the CO_2 emissions range results for full load (light green blocks) and MCL (thin green bars). They are very much in line with efficiency data.

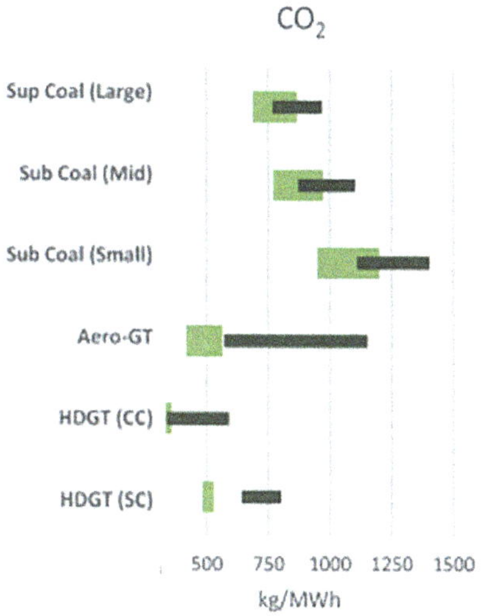

Two other considerations when variable inputs are present is the ramping up or down time, and the restart time.

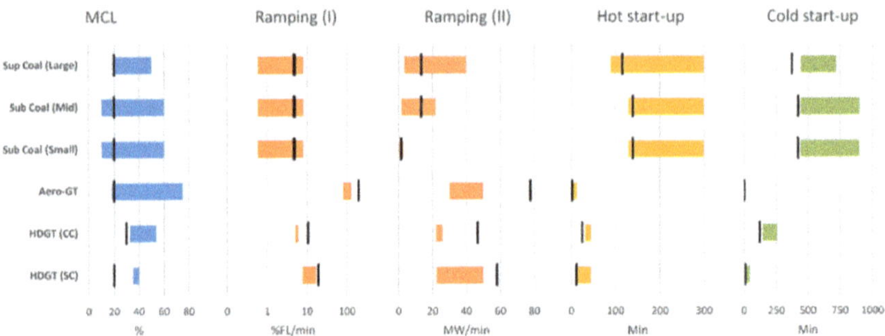

Not surprisingly, coal requires more time to adjust to changing conditions. It is a solid fuel and cannot be as easily dialed as a liquid or gas fuel. This is a negative aspect of coal when wind input is changing rapidly.

Another issue with wind turbines is that their manufacturing and installation requires large amounts of fossil fuels to accomplish. The main culprit is the production of steel and cement, and then the installation of large amounts of transmission lines needed to connect each windmill to the larger network. I have found many studies that assess a "payback" period, defined as the amount of operation time needed to recover the CO_2 emissions used in manufacturing. There is considerable variation on the predicted payback period to be found in the literature, ranging from 6 months to never. The expected life span is 20 years, at which time they need to be rebuilt. However, most of the materials are recyclable.

Assigning a payback period relies in the CO_2 emissions saving be accurately calculated during the period of operation, and as we saw above, that is a difficult task. Here I choose instead to evaluate emissions emitted during construction and installation of a turbine, and then this will be added as a debit in a calculation of net benefits that follows. Using data from a study (DOLAN, 2012) that reviewed numerous calculations on this subject, a "harmonized" amount of CO_2 emission was determined to be 2,365,200 kg to build and install a 3 MW turbine.

In a 2018 Harvard University study (Lee M. Miller, 2018), models show that when wind power is ramped up to very high levels in USA, (.46TW, the total demand in the country!), the turbines have important effect on vertical temperature gradients in the atmosphere, leading to a net warming effect over the whole USA of .24°C. Other commentators have read this paper and concluded that it is misleading: wind is not creating new heat; it is simply moving existing heat around. I have not been able to crack this technical nut myself, but it would not be surprising to me that slowing down the wind on a large scale would have some climatic effects. On the other hand, the study covers USA only, and that cannot be extrapolated to world temperature change considering that 2/3 of the planet is covered by ocean, where there will be very little installed wind power! Furthermore, it is acknowledged that wind needs back-up power, so assuming it provides 100% of demand is outside the bounds of reality.

Another Harvard study (Keith A. S., 2013) shows that very large wind farms, which would be necessary if we are to expect important contributions to emissions reduction, reach a saturation point due to the drag induced onto downstream turbines by the upstream ones.

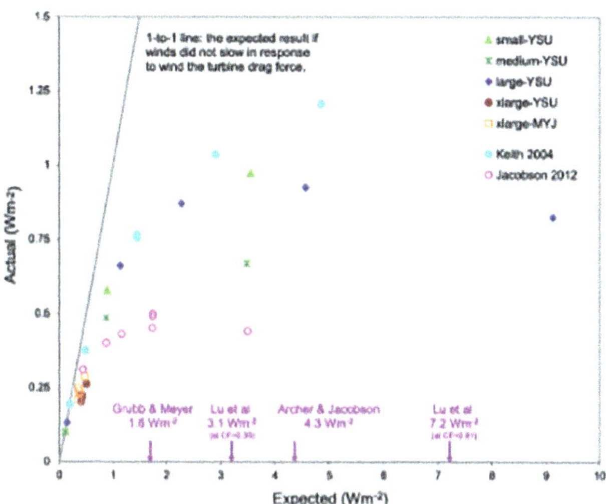

If output of individual turbines were unaffected by the size of the farm, it would follow the solid black line. The study on theoretical farms of different size show the output density tapering off as the farms get bigger. Each set of points follows a different path because there are also differences in assumed capacity factors in the reported studies. The message here is: wind turbines must be spaced away from each other so that additional drag is not induced. The authors quote a saturation point of around 1 W/m2. For example, an array of 100 turbines of 3 MW each would reach saturation if placed over an area of (100 x 3 x 1,000,000)/1 = 300,000,000 m^2, or 300 km^2. Any additional turbines added would have minimal effect. If we believe this calculation, then only 3 turbines could be placed in each square kilometers of area, assuming they are part of a large array. How large? The smallest area considered here is 2700 km^2, or an area of 52 x 52 km. The largest considers the whole planet (identified as Jacobsen 2012 in the chart).

Another study (S. C. Pryor, 2020) showed opposing results. It was assumed that installation of larger capacity turbines would be used to avoid competition for land. Quadrupling of current capacity produced a theoretical increase factor of 3.63 rather than 4 due to system-wide inefficiencies. There was no appreciable effect on temperature.

I am inclined for my purposes here to ignore the possibility of additional planetary warming being caused by the presence of wind turbines, but I have no doubt that large arrays of them will have negative effects on their net capacity density. I consider that it is unlikely that land will be available to create wind farms large enough to handle 30 to 50% of electricity demand, with sufficient spacing between them to avoid negative interaction. We will return to this issue after some calculations on how many turbines we need to get where we want to go.

ANALYTICAL STUDY: IMPACT OF LARGE GLOBAL WIND PENETRATION ON EMISSIONS

The information described above must seem like an overload, and it might be difficult to glean what it all means going forward. Here I make some gross assumptions in an analysis that attempts to answer the question: what can the impact be of a large amount of wind power on CO_2 emissions reduction? My assumptions are as follows:

1. I will study a gradual buildup of new turbines over a 20-year period, from 2020 to 2040.

2. Target penetration in 2040 is 30% of global electricity production that is currently provided by fossil fuels. Any number larger than this leaves little room for other technologies, and we know that fuel-powered plants that are required for back up become inefficient when they operate at small part power levels. Furthermore, a penetration of 30%, when combined with a typical capacity factor of 30%, would mean that with ideal wind conditions, wind would be temporarily be providing 100% of power demand. The current capacity for wind is around 2.5%, and I will consider the benefits of that in the calculation.

3. I will assume nominal capacity factor of 30%. UK studies show average 27% (R Camilla Thomson, 2015) and others, mainly from USA, show higher. An average of 30% is reasonable based on all the papers reviewed. However, keeping in mind that there is an expected decline in performance for large arrays of turbine, I will use 30% at the beginning of the study and 90% of 30% = 27% at the end. (90% is taken from the Pryor study.

4. CO_2 emissions for construction and installation will be added for each turbine, assuming that they apply over a 1-year construction period. I will use 2,365,200 kg of CO_2 per 3 MW unit, as described previously. This will tend to get offset by the savings being generated by the 2.5% of world capacity that is already operating.

5. Backup power is assumed to be supplied by coal and gas, in proportion to their current levels of usage: 62.4% coal and 37.6% gas (Report, 2020). As detailed in Chapter 4, it seems unlikely that coal can be replaced considering its abundant supply compared to natural gas. CO_2 emissions are then calculated based on the Lew and Gonzales–Salazara papers described above. I have combined coal and gas profiles to arrive at the following chart.

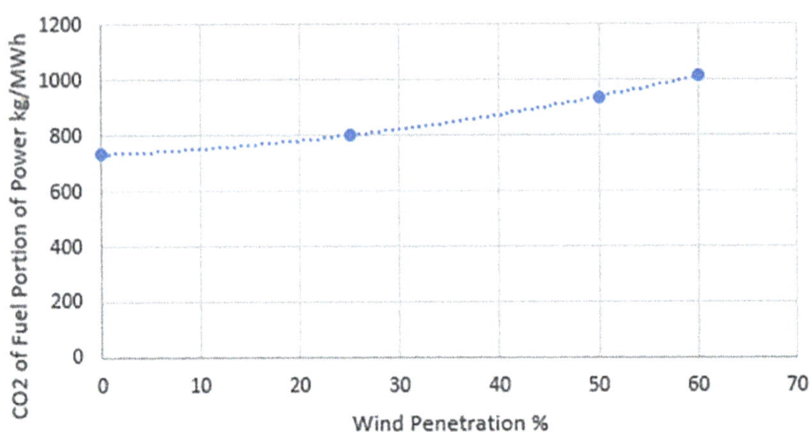

With this chart, we consider efficiency debit for power plants running at part power. The CO_2 density savings are then the given by this function.

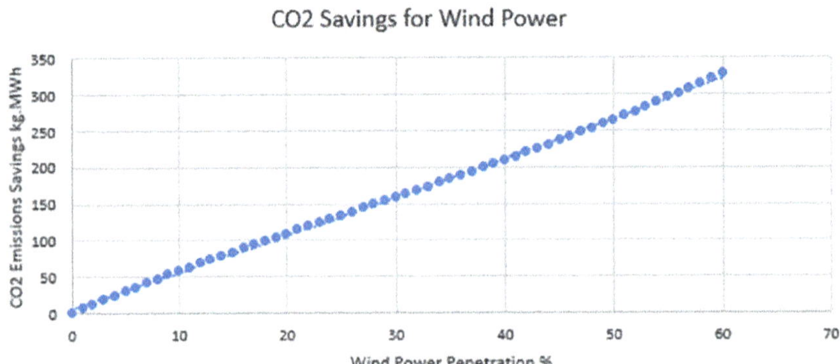

In reality, as discussed above, a turbine at 30% capacity factor can operate at anywhere from 0 to 100% demand at any one moment, and it therefore would follow all of this curve and beyond. I expect, however, that adjustments will be made operationally to limit the debits associated with wind at extremely high penetrations. I assume here that the globally averaged savings will follow this curve.

6. In order to achieve 30% penetration from today's 2.5%, we need to produce 27.5% of fuel-driven demand for electricity in 2041, which is projected to be 19505 TWh based on the assumptions on demand described earlier in this chapter. Considering a capacity factor of 28.5% average, and a turbine rating of 3 MW, this will require the construction of 781250 turbines, or 39062 per year over 20 years, or 3255 per month.

Here is the result of the study

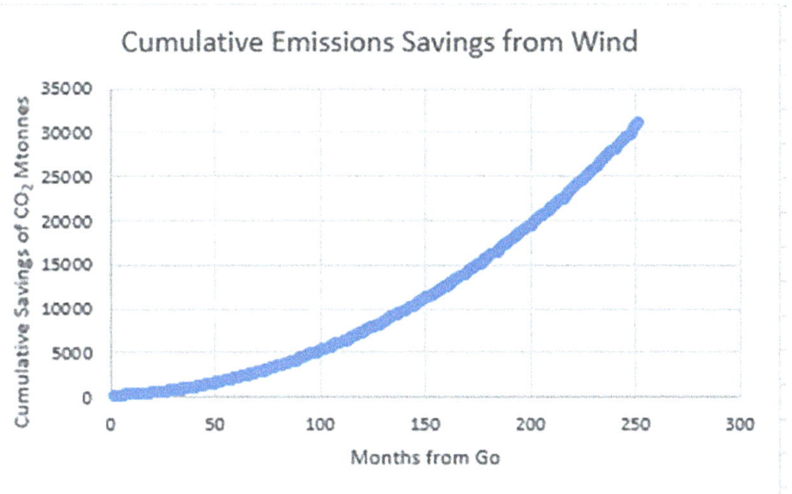

The chart shows slow buildup of savings at the beginning, reflecting low penetration and also the debits associated with building the turbines. Benefits increase over time up to the end of year 21. The additional year is added to give a chance for the turbines constructed in the 20th year to come online.

The cumulative savings of 30,965 Mtonnes of CO_2 needs to be translated to a saving in global temperature increase. I use the IPPC exchange factor previously discussed in Chapter 4 in the section describing the carbon budget to achieve particular level of temperature increase. The exchange factor is 1838.9 Gtonnes per °C. After 21 years of effort on building 781250 wind turbines, the benefit in temperature is .017°C.

This low number startled me, so I then extended the turbines effect to the end of the century. Prediction is difficult because all of the turbines would need to be decommissioned at some point and then rebuilt, recycling much of the same material. New ones would also need to be built to maintain 30% share with the energy demand constantly increasing. I made the following assumptions:

1. To maintain 30% share of demand, more turbines need to be built bringing the total to 1.07 Million.
2. Construction emissions for the extra new turbines are same as before
3. Emissions savings are considered to be linearly increasing from 2041 value to 2100 value.
4. No new emissions are considered for reconstruction of decommissioned turbines

The result shows that net emissions savings reach 318,991 Mtonnes, equivalent to a temperature saving of .17°C. Here is the extended emissions function.

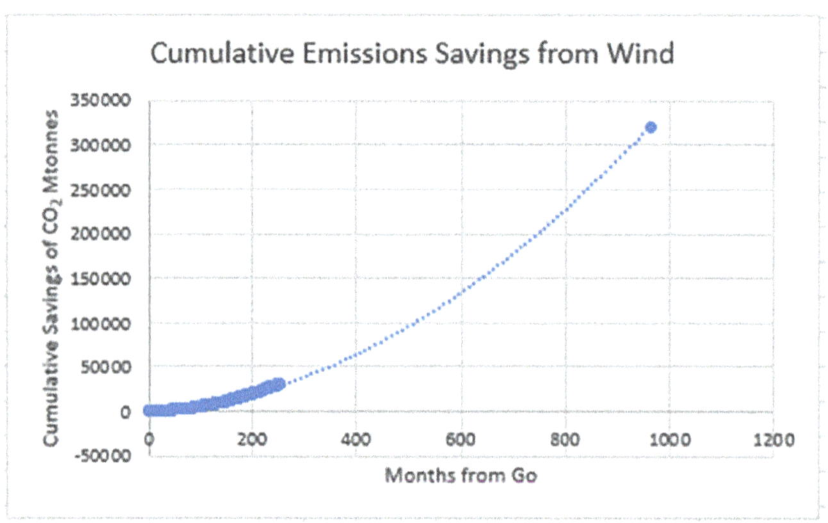

The relatively small amount of savings in temperature is generated by the IPCC exchange factor of 1838 Gtonnes/°C. Here is a table on the carbon budget from IPCC report on mitigation plans to achieve 1.5°C or 2°C temperature increase since pre-industrial times.

Additional Warming since 2006–2015 [°C][1]	Approximate Warming since 1850–1900 [°C][3]	Remaining Carbon Budget (Excluding Additional Earth System Feedbacks[3]) [GtCO$_2$ from 1.1.2018][2]			Key Uncertainties and Variations[4]					
		Percentiles of TCRE [3]			Earth System Feedbacks [5]	Non-CO$_2$ scenario variation [6]	Non-CO$_2$ forcing and response uncertainty	TCRE distribution uncertainty [7]	Historical temperature uncertainty [1]	Recent emissions uncertainty [8]
		33rd	50th	67th	[GtCO$_2$]	[GtCO$_2$]	[GtCO$_2$]	[GtCO$_2$]	[GtCO$_2$]	[GtCO$_2$]
0.3		290	160	80	Budgets on the left are reduced by about −100 on centennial time scales	±250	−400 to +200	+100 to +200	±250	±20
0.4		530	350	230						
0.5		770	530	380						
0.53	~1.5°C	840	580	420						
0.6		1010	710	530						
0.63		1080	770	570						
0.7		1240	900	680						
0.78		1440	1040	800						
0.8		1480	1080	830						
0.9		1720	1260	980						
1		1960	1450	1130						
1.03	~2°C	2030	1500	1170						
1.1		2200	1630	1280						
1.13		2270	1690	1320						
1.2		2440	1820	1430						

I have graphed this data for the 50th percentile in the analysis uncertainty range.

The slope of this line is 1838.9 Gtonnes per °C, and this is where my exchange factor comes from. Note that in the table there are assumptions about the effects of other non-CO2 greenhouse gases, and any future changes there would naturally change the CO$_2$ portion also.

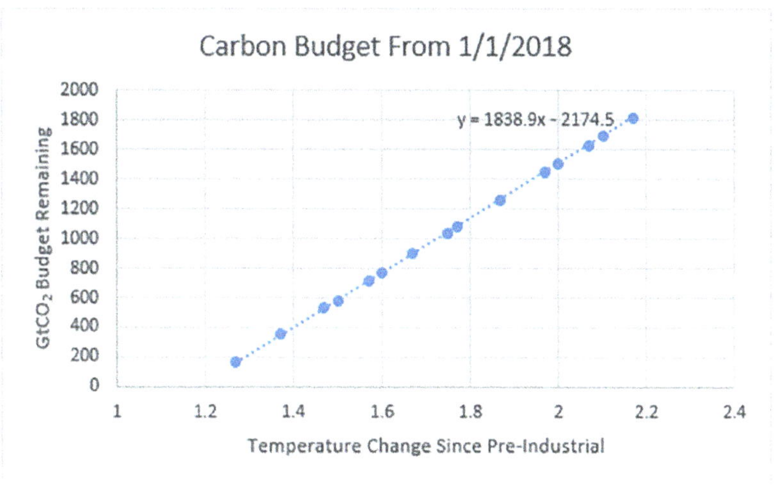

I did my own calculations for this exchange factor. From Chapter 1, CO_2 concentration in the atmosphere went from 225 to 400 ppm since 1880, and global average temperature increase was 1.25°C. The increase of 175 ppm is equivalent to 901 Gtonnes of CO_2. The exchange rate would then be 721 Gtonnes/°C. This is smaller than the IPCC value because about half of the CO2 released is absorbed by the oceans, plants, and trees. It does not all go into warming of the planet. If the Earth's absorption percentage is 50%, then the exchange factor would be 1422 Gtonnes/°C. Therefore, I conclude that the IPCC exchange value of 1838.9 is reasonable, given that their calculation must certainly be more complete than mine. However, it relies on plants and the oceans continuing to absorb more and more CO_2.

The benefits of such a massive project, where more than 1 Million turbines are constructed and put into service, seem small. This must be an indication that efforts much greater than this will be needed to make real progress.

SOLAR POWER

Solar power is also an intermittent source of energy due to varying cloud cover. An additional disadvantage is that the efficiency of energy extraction is also dependent on latitude. At higher latitudes, the sun's average intensity is reduced as compared to the tropics. The Earth's wobble brings northern and southern areas into regions of temporary higher efficiency during their respective summers, and lower efficiency in their winters.

There are two ways to get electricity from the sun. The conventional method is to use photovoltaic (PV) cells which convert light to direct current. This is how flat solar panels work. Another way is to use parabolic mirrors that create focused beams of heat, which in turn creates steam to drive a conventional steam turbine. This is called Concentrated Solar Power (CSP). The International Energy Agency (IEA) predicted in 2014 that solar could supply 27% of global electric power by 2050, 16% from PV and 11% from CSP. There is certainly a large year-over-year percentage increase in solar power, and it is reported to be arriving at lower and lower cost.

The largest PV generating station in the world in Kamataka, India, with a capacity of 2050 MW. This is equivalent to 683 wind turbines of 3 MW capacity. Asian countries are the biggest users so far, especially China, where most of the solar panels are made.

For CSP, there is a 392 MW facility in Nevada called the Ivanpah Solar Electric Generating System. It employs many thousands of individual flat mirrors that direct the sun's heat to towers containing boilers that create steam for electricity generation. The region is a desert that has a much higher-than-average availability of sunlight.

An interesting application for solar panels is local installation onto rooftops that have no other function (except to keep the rain out). It could be on private homes or large commercial or industrial buildings. Connecting all of these to the grid is complex, but there is also opportunity to use the power locally without grid connection, an interesting option for large cities hungry for power. Before becoming too excited by the idea, most building occupants need some level of continuous supply, and solar cannot promise that. But with some innovation it could certainly provide partial supply.

Solar energy can be extracted at higher intensities compared to wind when it is considered from the point of view of required land area, which is certainly an important consideration if we want a greatly increased contribution of solar power. Here is a chart comparing wind and solar power densities in the USA. (Keith L. M., 2018)

Legend: Red & Yellow = Solar   Blues = Wind

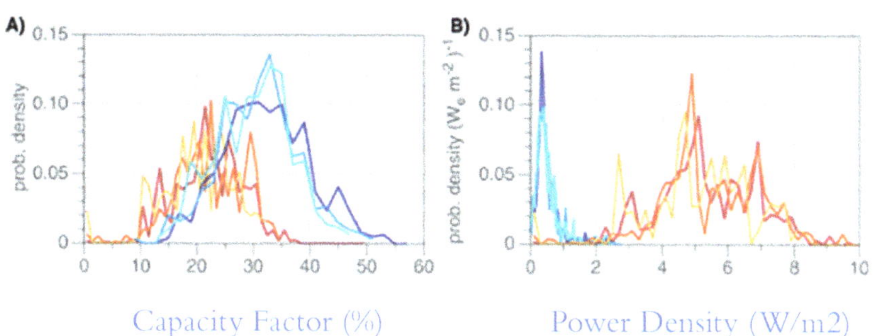

Capacity factors for solar are lower than for wind, primarily because there is no sun at night. However, power densities are much higher for solar. It takes space to place wind turbines in such a way that they do not interfere with each other, whereas solar panels can fill an assigned area almost entirely.

The USA would be considered a prime candidate for solar power. Latitude range of 30° to 45° is fairly low, indicating high solar energy density. Land availability should be good, considering population density is low, especially in the western half of the country. Yet solar power accounts for only 1.8% of US electricity generation, compared to 6% for wind. Land use policies may be affecting this result. Wind turbines can be placed on active farmlands without any significant effect on farm productivity. The same cannot be said for solar, which would evidently rob the crops of their sole source of energy. Mountainous and forested regions might be unsuitable for either wind or solar.

The Keith paper on power density makes an interesting observation. Considering the observed power densities and the theoretical scenario where either wind or solar could provide 100% of US energy needs, land use for wind would be 72% of the continental US, and solar would be 6%. This scenario is evidently not realistic because it is obvious that a mix of power sources are needed, but it still illustrates that wind power is considerably more unlikely to achieve a prominent position, while solar has more of a chance.

CO_2 intensity for manufacturing and installing solar receptors is considerably less than for wind turbines. They also have longer expected life before needing replacement. It would seem, then, that solar is a better alternative than wind. On the other hand, wind may be more useful in far north or far south regions where solar intensity is weak.

The cost of solar power is the most probable cause of wind power becoming more prevalent. However, recently the cost has come down dramatically.

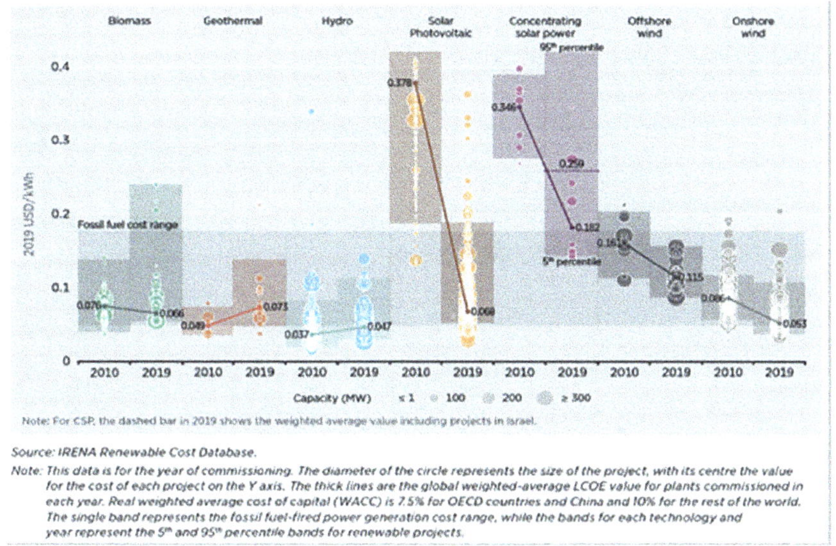

Note: For CSP, the dashed bar in 2019 shows the weighted average value including projects in Israel.

Source: IRENA Renewable Cost Database.

Note: This data is for the year of commissioning. The diameter of the circle represents the size of the project, with its centre the value for the cost of each project on the Y axis. The thick lines are the global weighted-average LCOE value for plants commissioned in each year. Real weighted average cost of capital (WACC) is 7.5% for OECD countries and China and 10% for the rest of the world. The single band represents the fossil fuel-fired power generation cost range, while the bands for each technology and year represent the 5th and 95th percentile bands for renewable projects.

This development is likely to lead to greater and greater renewable market share for solar power.

A key issue is the possibility of excess energy storage during sunny times so that loading can be levelled. Large Utility-level batteries have been studied for a 4-hour operating period to fill gaps in solar energy production. The cost of the batteries is similar to the cost of the PV panels that provide the energy, so this roughly doubles the cost of using solar energy.

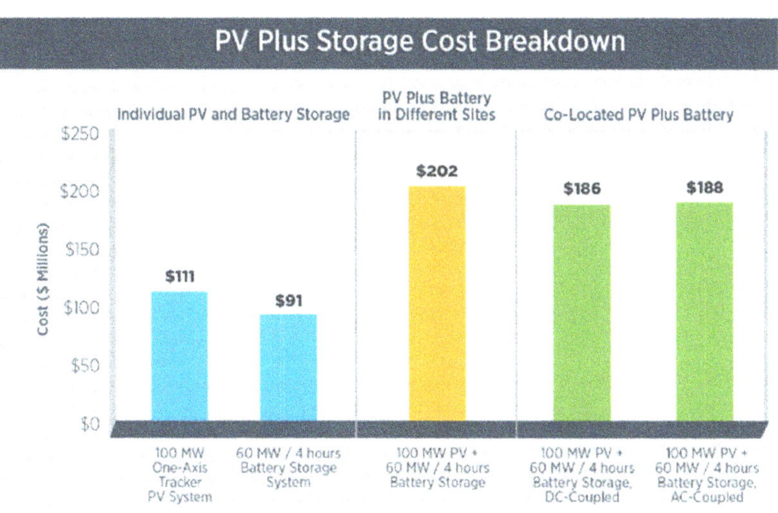

I was curious about the weight of these batteries. Here is a chart showing the reason why Lithium Ion batteries have taken hold.

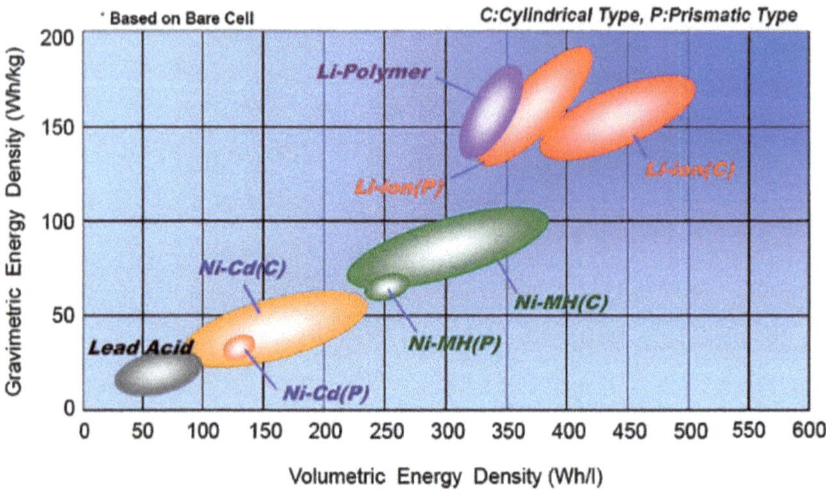

Both the volume and weight are significantly reduced as compared to other battery types. Still, I wondered what a 60 MW battery that is meant to supply power for 4 hours would weigh. Using 150 Wh/kg trade factor, the answer is 1,600,000 kg, or 1.6 Mtonnes. Note that this is only the weight of the battery cells and does not include all of the other equipment needed to make it work. If we go back to the IEA target of 16% of global electricity demand provided by PV systems by 2050, a total delivered energy of 4264 TWh would be required, assuming demand is unchanged from 2019. If we further assume that this power is all backed up by batteries, and the batteries operate 4 hours per day out of 24, then batteries need to supply 710 TWh of energy. We would need 2.96 million batteries weighing 1.6 Mtonnes each, or 4.73 Ttonnes (trillions of tonnes). Does this make sense? Current estimate of world Lithium supply available, including those as yet unexploited, is 65Mtonnes. A lithium-ion battery has approximately 2.3% of lithium, so the total weight of battery that can be made is 2826 Mtonnes, or .0028 Ttonnes, a tiny fraction of the "demand" described above. I conclude that rolling out a broad system of solar power backed up by batteries is not feasible. It could perhaps be used for niche applications.

I add here that the cost of battery backup expressed in the storage cost breakdown chart above undoubtedly uses battery costs at current or near-term projected rates. No account is taken of the fact that such large batteries would deplete the lithium supply, driving prices up.

All of this puts solar into the same category as wind power. It can supply a small fraction of demand, given that it has a capacity factor of only 20%, which requires large back-up power. Furthermore, it is unlikely that current battery technology could provide that back-up, so we once again revert to fossil fuels.

The opportunity to have local temporary power supplied to buildings with PV panels on their roofs could be explored. In October 2020, the State of South Australia announced that for about an hour on one day, 100% of the state's electricity needs were supplied by solar power. Interestingly, 73% of that was supplied by personal rooftop panels that are connected to the grid. At certain times, too much power was input making the grid unstable. That seems like a temporary solvable problem. Still, coal-fired power plants needed to redirect some of their output to neighboring states. I presume it was necessary to keep them running as a backdrop to a sudden loss of solar capacity.

I will not do any calculation of potential benefit as I did with wind power. The two technologies are similar in that they are variable and require back-up. Instead, I will generously assume that wind and solar together can provide 40% of electricity that would otherwise be provided by fossil fuels. The benefit of .17°C calculated for 30% wind can be extrapolated to .17 x 4/3 = .23°C for 40% total contribution. I will add an extra .01°C benefit for solar considering that emissions in the manufacturing stage are less than for wind turbines. The net benefit for wind and solar together is then .24°C, to be realized by year 2100. This will be revisited when I do a more comprehensive calculation for all energy sources in Chapter 7.

A REAL-WORLD STUDY FOR WIND AND SOLAR

Before moving on, we could look for countries that are already ahead of the game and get a real-world example of how emissions have declined when a large percentage of electricity is generated from clean, renewable sources. Germany is such a country. Energy is claimed to be 46% from renewable sources in 2019. This includes wind, solar and biomass. The penetration for wind and solar is 34%, and this is the relevant number for this study because burning biomass also produces CO2. The following 2 charts show the results (Roser, 2017).

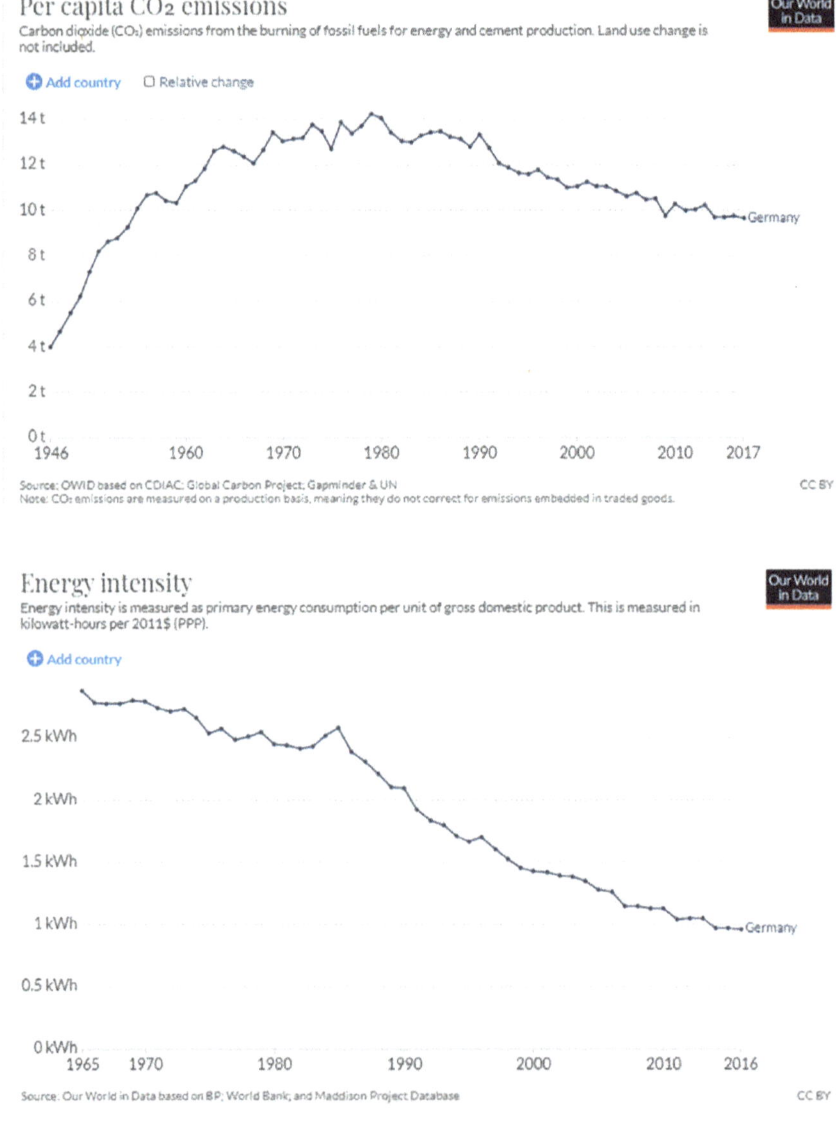

In the first chart, we see a laudable reduction in emissions starting all the way back in 1980. Can this be caused by incorporation of renewables? The second chart shows energy intensity as a portion of GDP. This shows that there has been a significant improvement in energy use, regardless of its source, going from 2.5 KWh to 1 KWh per unit of GDP. The 60% improvement in energy efficiency greatly exceeds the emissions reduction of 29%, so it is difficult to judge whether renewables have had a significant impact. Improvement in energy efficiency equates to a reduction in demand, which is a different category of savings. Therefore, I include a 3rd chart below.

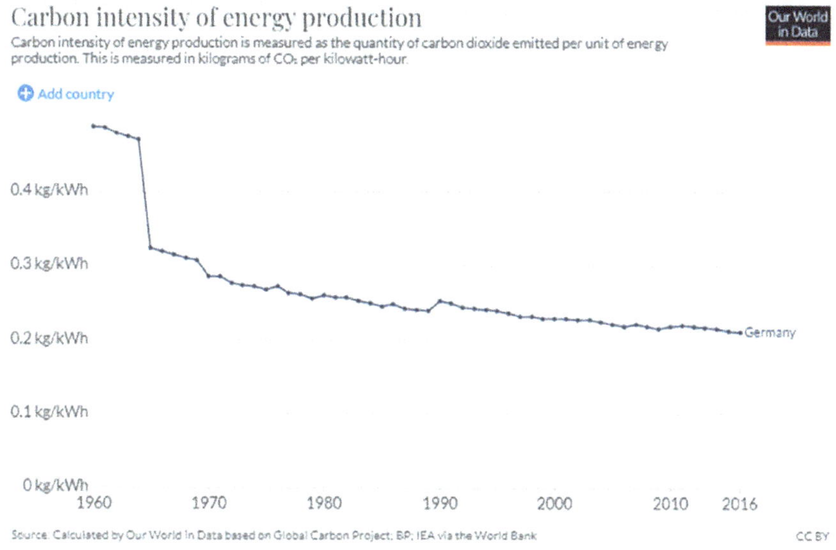

Since 1980, carbon intensity of energy production has decreased by 19%. The impact in electricity generation would be higher than 19%, because the chart includes carbon intensity of the transport sector, which is likely relatively unchanged. It should be noted that Germany is also at somewhat of a disadvantage here, because in this same time period they have added capacity in coal-based energy production as they have phased out "cleaner" nuclear power.

The conclusion is that, in this one example, an emissions reduction of 19% could be achieved with a 34% penetration of solar and wind for electricity. Does this make sense when compared to the small studies I made in the previous sections of this chapter? I looked at year 2100 in my wind study, where 30% penetration is achieved. I find the savings in emissions in that year was 15.5% compared to what would have been seen with only fossil fuels. This all seems to be in the same ballpark.

ELECTRIC VEHICLES (EV)

As with many other topics that I explore, I find wildly varying scenarios regarding the net benefit of EV's. It has been widely reported that the energy used to manufacture EV batteries, when added to the energy needed to constantly recharge them, results in zero benefit. Closer inspection reveals that the benefit range is indeed wide, depending on the fuel used to produce the batteries and to charge them.

I found a credible study on this subject (Lutsey, 2018) by the International Council on Clean Transportation (ICCT). This study evaluates the entire life-cycle emissions of EV's compared to gas-powered cars. In the chart below, here are the assumptions. "Fuel cycle" indicates all aspects of the production and transport of fuels used in the processes, as well as fuels used in electricity generation for recharging batteries. The EV considered in this study is the Nissan Leaf with a 30 KWh battery. The vertical lines superimposed onto the Lithium Battery bars indicate a range of emissions values for battery manufacturing that can be found in the literature. The particular value used here (green bar) assumes the batteries are made in Asia, where carbon-inefficient coal is the prime source of electricity generation used in the construction of the battery. It is assumed here that all cars, including the lithium batteries, will last the full life of the vehicle, considered to be 150,000 kms. Finally, I note that the tailpipe emissions for gas-powered cars in Europe are quite lower than those in North America due to more stringent emissions standards. A chart for North America would show higher tailpipe emissions. On the other hand, countries in Asia and Africa tend to use small vehicles, meaning their emissions would be lower.

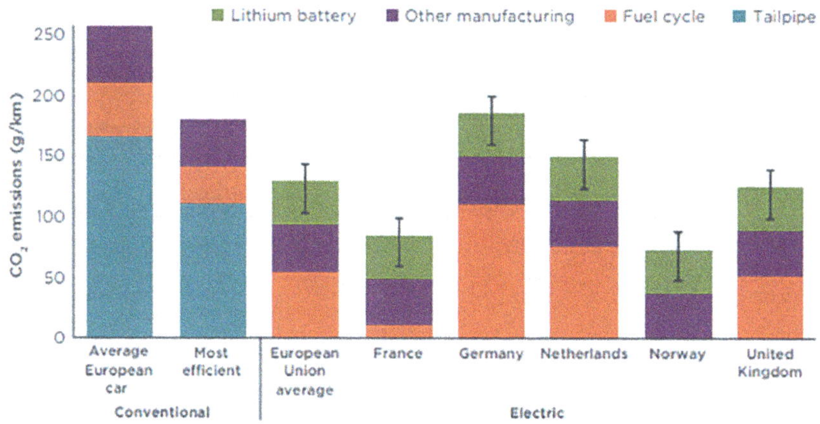

The benefits depend very much on the source of electricity in the particular region of interest. In Germany, where coal is an important source of electricity, a small EV shows no benefit over the most efficient gas-powered vehicle. In France, where prime electricity source is nuclear, the benefits are substantial.

This study uses the Nissan Leaf with a 30KWh battery. Its range is reported to be only 107 kms, so this is unlikely to have a sustained interest in the marketplace. For example, a Tesla Model S has an 85KWh battery and its range is much more practical. However, such a battery could make the green bar above around 3 times the height. The purple bar representing other manufacturing-related emissions would likely be around 1.5 times because it is a much larger, heavier vehicle than the Leaf. It is obvious without drawing it that EV's would not show much advantage except in countries where the electricity is generated from low emission sources like nuclear and hydro.

This study assumes manufacturing of the battery in Asia with coal as the energy source. Tesla makes its own batteries in Nevada, and they are adding a huge amount of solar power to their plant. In addition, back-up power in the state of Nevada is already 30% renewable, and the rest is provided by more carbon-efficient natural gas (as compared to coal in Asia). This means that the manufacturing of a Tesla battery will be more efficient from an emissions point of view than those made in Asia. Still, the portion of battery-related emissions that comes from the mining of the material resources will be high. Furthermore, Tesla still makes only a small fraction of EV batteries, most are made in Asia. It seems unlikely that Tesla can capture a large part of the world market considering that the cost of their vehicles is out of reach of most buyers in the world.

Let us return to our calculation on lithium supply. How many 85KWh batteries could be made with the current known supply of lithium:

Output of Lithium-Ion Battery	150 Wh/kg
Weight of 85 KWh battery	(85 x 1000)/150 = 567 kg
Fraction Attributed to Lithium	.023
Lithium Content	.023 x 567 = 13.03 kg
Current Lithium Reserves	65 Mtonnes = 65,000,000,000 kgs
Number of Vehicles	65,000,000,000/13.03 = 4,988,488,104

The result is around 5 billion cars. To put this in perspective, there are currently about 1.4 billion passenger cars in service, and the 2019 annual rate of production is 70 million. It looks like there is plenty of lithium for making EV batteries for the foreseeable future.

But should we? Can we keep making personal use vehicles that weigh 1500 kg just to carry us around? If electric cars take hold, then we will be making 2000 kg cars just to carry us around. Is it possible that the "green" value of electric vehicles will make us believe that we can keep making so many cars without having to make any personal sacrifices?

Let us look at the total emissions of all of these cars. We will assume that the chart above is reasonably accurate: 250g/km travelled and 150,000 kms average usage. This gives 37500 kg of CO_2 per vehicle. I will assume car production increases at the same rate as the increase in energy demand already discussed. This means that by year 2100 car production will reach 108,080,000 cars. The emissions from all these cars over 80 years will be 267.1 Gtonnes. This translates to a global temperature increase of .15°C using the exchange factor of 1838.9 Gtonnes/°C.

If we generously assume that the 107 KWh Nissan Leaf battery is adequate for European countries, then the chart from the Lutsey study shows that an average of around 40% savings in emissions can be achieved. In the rest of the world, where coal dominates, it would be considerably less. Furthermore, larger EV batteries are needed, making the savings even less. I will assume 25% savings if 100% of cars are converted to EV's. Realistically, though, it is unlikely that a market share of more than 50% can be reached. Therefore, the savings would be .15°C x 25% x 50% = .02°C.

This amount is so low that I feel that the quality of my assumptions in these calculations is unimportant. If I were off by a factor of 3, it still leaves only .06°C effect of EV's by year 2100.

It appears that cars in general are not as strong a factor in global warming as most people think. Can this be true? Here is a breakdown of where all emissions are coming from. The data is from 2010 and is contained in the 2014 IPCC reports.

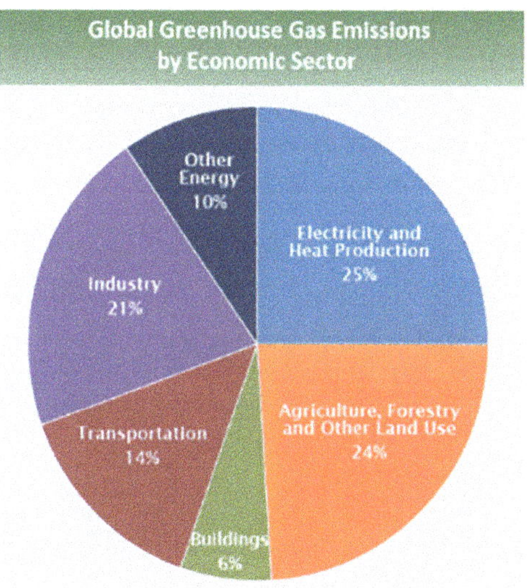

Transportation accounts for 14% of all emissions. This includes emissions from heavy trucks and aircraft. The contribution from passenger vehicles is about 6%. Considering a global temperature increase of around 1°C, then passenger cars might be responsible for about .06°C. This means that my simple calculations for future savings in temperature are in the right ballpark.

There are still regions of the world where EV's will make sense for local Climate Change action plans. For example, in the Canadian province of Quebec, electricity is almost exclusively provided by a wealth of hydroelectric facilities, which create minimal GHG's. Therefore, contributions to GHG emissions are more highly loaded towards transport. Personal vehicles account for about 20% of GHG emissions in Quebec. If these vehicles can use clean electricity to charge them, then it might make sense to do so.

ELECTRIC TRUCKS

I would like to now consider the other main contributor to transport emissions, large trucks. Can these be made fully electric?

Here we have to consider the fraction of total vehicle weight that is dedicated to payload. A 1500 kg passenger car can typically carry a payload of around 550 kgs, equivalent to 4 or 5 people with their belongings. The addition of a 500 kg battery makes the vehicle heavier, and it needs to be reinforced to account for it. Still, it is not unreasonable to do so.

An 18-wheel transport truck has a vastly different ratio. Empty weight, including a trailer, is around 15,900 kg. Maximum cargo weight is 20500 kg. The "payload" is considerably more than the vehicle weight. Right away, we can expect that the addition of a heavy battery to drag all this weight around might be prohibitive but let us find out.

If the size of the battery is proportional to the total vehicle weight, then the weight of the battery for an 18-wheeler would be calculated as follows:

Passenger vehicle battery weight Fraction: 500/2000 = .25

18-wheeler battery weight: .25 x (15900+20500) = 9100 kg

This means that we need to add 9100 kg to the weight of the truck, thus reducing its payload to 20500 − 9100 = 11,400 kg. The payload is half, and the range will be greatly reduced for vehicles that by nature are built for long haul trips.

I conclude that with current battery technology it would be nonsensical to produce large battery powered trucks. It is likely that we will start to see pick-up trucks in EV form, but these are more like passenger vehicles in the consumer world.

BIOMASS

Biomass energy refers almost exclusively to the burning of organic material to generate heat, which is then used directly, or, more commonly, to generate steam to drive turbines that generate electricity. All organic material is carbon-based, so naturally CO_2 is generated. The main advantage is that, unlike fossil fuel, it is renewable in the sense that it can be regrown. Carbon emissions in some cases can be somewhat mitigated by carbon capture by the plants themselves, especially when they are perennials that capture carbon in the soil and in their root systems. Furthermore, the burning of waste that would otherwise go into landfill can avoid methane emissions from anaerobic decomposition.

Biomass is commonly used in developing countries for local heating and cooking. The simplest form is a wood fire. Such fires burn very inefficiently, and they create not only CO_2 but also particulate emissions that are directly harmful to humans. Much of the air pollution in some countries is caused by the burning of wood or other fuels for cooking, or by the burning of residual matter on farms in preparation for replanting. This kind of combustion is considered in the IPPC's accounting of global CO_2 emissions.

I have not been able to arrive at a sensible accounting of any savings in GHG emissions from biomass. Certainly, worldwide it is a net CO_2 emitter. However dedicated power plants which recycle waste and collect crop residues may well produce a benefit.

BIOFUELS

Ethanol and biodiesel fuels are created from crops such as corn and wheat. These fuels are added to gasoline in small percentages. They have lower energy value that fossil-fuel gasoline, but they increase oxygenation of the fuel, allowing for slightly better fuel consumption and reduced pollution.

These biofuels require considerable energy to produce, and where energy is provided by fossil fuel-driven power plants, the net gain for biofuels is questionable. I have found several papers that conclude that energy out is more than energy in, leading to the assumed questionable conclusion that it is useful to pursue biofuels as an energy source. The ratio of energy out to energy in the USA using corn as a source is about 1.5:1, while gasoline from fossil fuels is at about 11:1. It is not obvious that biofuels are a sensible choice.

A second issue is the use of viable farmland to grow crops for fuel instead of for food. Such decisions may be driven by economics rather than real need. There is an extraordinarily strong lobby in the USA for biofuel producers who certainly have a strong steady income from a source that is mandated by law to be a percentage of all gasoline sold. This does not mean it is the right thing to do.

I expect that if fossil fuels become too scarce, as hypothesized in Chapter 4, then biofuels will become more interesting.

Certain fuels produced from biomass have been classified as sustainable. For example, Sustainable Aviation Fuels (SAF) can be so classified if it can be shown that they are carbon neutral on a life cycle basis. This means that the biomass from which they are produced must be a sustainable crop that absorbs as much CO_2 and the fuel will emit when burned.

For the purposes of this book, I will declare biofuels to be carbon neutral and not explore them any further.

REDUCTION IN CONSUMPTION

In all of the economies of the world, and especially in those considered to be capitalist, we rely fundamentally on economic growth. This drives the corporate profit motive which in turn drives stock markets all over the world. It relies on the size of the pie getting bigger and bigger so that even if one entity keeps the same percentage of the pie, that entity is growing. In recent decades massive amounts of immigration from poor to rich nations has been driven less by altruistic motives and more by the need for more workers and more consumers.

This means that consumption, on average, must go up. In the context of this book, that means consumption of food and energy must go up. How can this trend be reversed?

I do not have any new insights into how to make this happen. I especially cannot see how it can be done without shrinking economies in high consumption nations.

Returning to the German study on effects of wind and solar, there was a 19% reduction in CO_2 intensity, and a 60% reduction in energy intensity as a percentage of GDP. We can conclude then that while real world incorporation of wind and solar has been a useful and laudable effort, reduction in energy consumption is by far the most significant achievement.

Reduction in consumption has two components: improved efficiency and reduced energy demand. Examples of improved efficiency are switching from incandescent to LED lights and improved thermal efficiency of combustion processes. Examples of reduced demand are turning the heat down in the house and buying a smaller car to save fuel.

Below is the energy consumption chart for the entire world. The scale is in Megajoules per $ of GDP.

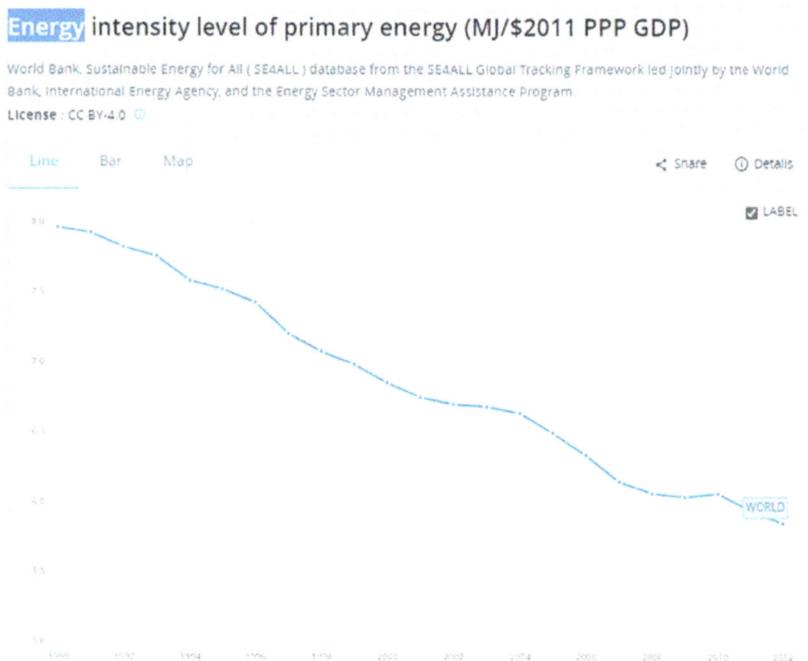

We see a 27% reduction between the years 1990 and 2012. This is encouraging but Germany has certainly shown that much more can be done, where they achieved 60% reduction in the same time span.

Regarding improvements in energy efficiency, the world is already on board to a great extent. In every aspect of the economy, the awareness of efficiency is high and the drive to improve is strong. I can validate one aspect of this from my own professional experience. For many years I was employed by a company that designs and produces gas turbine engines for airplanes and helicopters. There has been a consistent march towards improved fuel efficiency for decades. The reasons may include the wish for reduced emissions, but the truth is that the main driver is operating cost for the aircraft owners. In spite of the high cost of airplanes, it turns out that fuel charges are by far the highest total cost that an airplane will incur over its lifespan. A few percentage points of efficiency make the difference between a good product and a winning product in the marketplace. In the automotive world, total fuel cost can also approach the initial price of the car, and the majority of consumers consider fuel efficiency carefully before making a purchase. From home appliances to lighting to building insulation, there has been a relentless march towards greater efficiency, and it is paying off.

On the other hand, reducing energy demand has been a more challenging task. Demand has been increasing continuously. We have to conclude that the improvements in energy intensity (that is, energy per unit of GDP) have been achieved mostly through efficiency gains rather than reductions in energy demand. Perhaps this is the area where substantial future gains can be made.

METHANE AND NITROUS OXIDE REDUCTION

I will not elaborate on the subject on Non-CO2 emissions in this chapter, primarily because efforts to reduce them are so far quite minimal. In fact, the important efforts are still in the exploration and policy-setting phases, rather than implementation phase. However, it is extremely important that it be addressed, and as such I will elaborate more in Chapter 7 where we consider future strategies for mitigation of GHG's.

OTHER SOURCES OF GREENHOUSE GAS EMISSIONS REDUCTIONS

To summarize, by my very rough calculations, the total savings from wind, solar and electric passenger cars is .26°C by the time we reach year 2100.

This leaves us with the following questions: What are all of those commitments being made by signatories to the Paris Accord? Where are the savings coming from?

Recall that the pathway defined as having no mitigation to global warming is projected to see a rise in temperature of around 4°C by 2100 (RCP8.5), while the path with identified mitigation plans leads to 2°C I (RCP6.0). Therefore, we are looking for plans that takes the difference, 2°C, out of the atmosphere. So far, we have found a possible .26°C with wind, solar and electric vehicles. Here is a list of other heavy hitters that appear in the plans of many countries as part of the Paris accord.

Reduction in Deforestation

Reforestation

Improved Land Use as a Carbon Capture Mechanism

Decommissioning of Coal-Fired Power Plants

Improved Energy Efficiencies to Drive Down Demand

Carbon Pricing to Drive Down Demand for Fossil Fuels

Specific Reductions in Fugitive Emissions from Fuel Extraction Sites and Power Plants

Mitigation of N_2O and CH_4 emissions through improved farming practices

Carbon Sequestration

I will address all of these and other future opportunities in Chapter 7, where I attempt to make some specific recommendations going forward.

CONCLUSIONS FROM THIS CHAPTER

The solutions for GHG reductions that are most commonly in the public eye are relatively weak when scrutinized with some simple calculations.

Of the roughly 2°C reduction in temperature escalation that we are looking for, wind and solar power cannot likely provide more than .24°C, and this with a dramatic investment in infrastructure from now until year 2100.

Electric Vehicles are even less interesting, providing a possible attenuation of only .02°C with 50% market share.

Large transport trucks are not candidates for electrification because a significant part of their payload would be reduced by the heavy batteries needed to drive them.

Biomass produces its own CO_2 emissions and accordingly it will require careful management to provide any benefit. Burning of biomass also produces N_2O emissions which are very harmful.

Biofuels require large amount of energy to produce, and they displace valuable farmland that might better be used for food production.

Significant global reduction in energy demand is likely to be needed, separate from the reductions produced by improved process efficiencies.

Other solutions will be explored in Chapter 7.

CHAPTER 6

THE GEOPOLITICAL LANDSCAPE

A CANADIAN PERSPECTIVE

I WILL START THIS CHAPTER with a collection of anecdotes from my home country, Canada.

In 2000 a minister in Canada's ruling Liberal party by the name of Stephane Dion made a splash by successfully drafting and passing a law called "The Clarity Act". This law preemptively addressed separatist aspirations in the province of Quebec by enforcing 2 simple rules: a referendum on separation must be accompanied by a question that is clear in its description of what is being proposed, and separation must only be moved forward if there is a clear majority expressing consent. A clear majority was meant to imply that a simple 50% plus one result is insufficient to break up a country. If such a sensible law had been passed in Great Britain, then Brexit would not have come to be.

In 2006 Mr. Dion ran for and successfully obtained the Liberal party leader position in preparation for a planned 2008 election. Public perception of him was good. He was seen as a man of principle and intelligence, without any obvious political biases. Then, in 2008, he announced as part of the party platform a climate change plan with the primary element being a carbon tax. Oops! The rival Conservative party jumped on it, making it clear to the Canadian public that the price of everything would be going up.

Both parties knew from polling and surveys that the average Canadian voter wanted their leaders to sign up to climate change action, but perhaps not to the extent that any plan would dig into their pockets. The Liberals countered this with a plan to reduce income taxes to compensate for carbon taxes, but that subtlety was lost on the public. This is especially true because of Canada's severely graduated tax rates, where the bottom 20% of income earners pay only 0.6% of income tax, and the top 20% earners pay 67% of all income taxes. (Data from the Fraser Institute).

The Liberals lost the election. There were several reasons for it, but undoubtedly their "Green Shift" plan and the carbon tax made a hefty contribution to the loss. The Conservative party won the election and later withdrew Canada from the Kyoto Protocol.

Today, the Liberals are back in power, and the carbon tax has returned, at least in spirit. The Trudeau government has a new spin on it. Don't worry about the tax, because whatever we take from you, we'll give it right back in rebates! I'm not kidding. It is mystifying how demand is expected to decrease when there is no real change in price. Of course, there will be no rebates for businesses, only consumers, who happen to be the voters.

Several Canadian provinces have gone ahead with their own variations on carbon taxes. On the other hand, other provinces are currently challenging the federal carbon taxes in court. I am not sure if anything will ever really happen.

Let us say it does happen. Will carbon taxes work in reducing demand for fossil fuels? The two charts below might give us a clue.

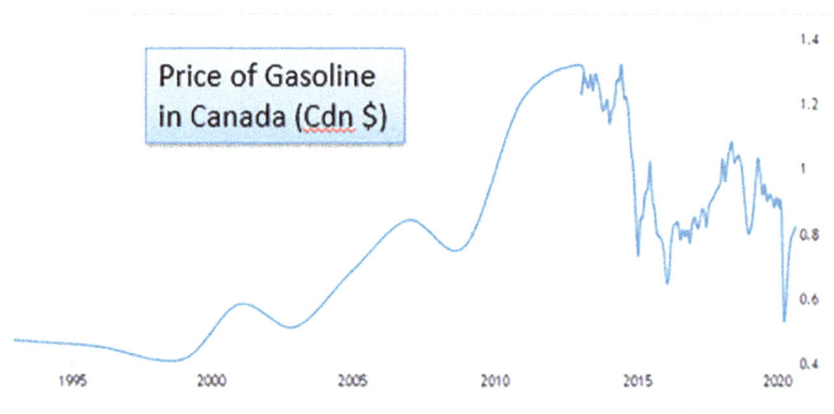

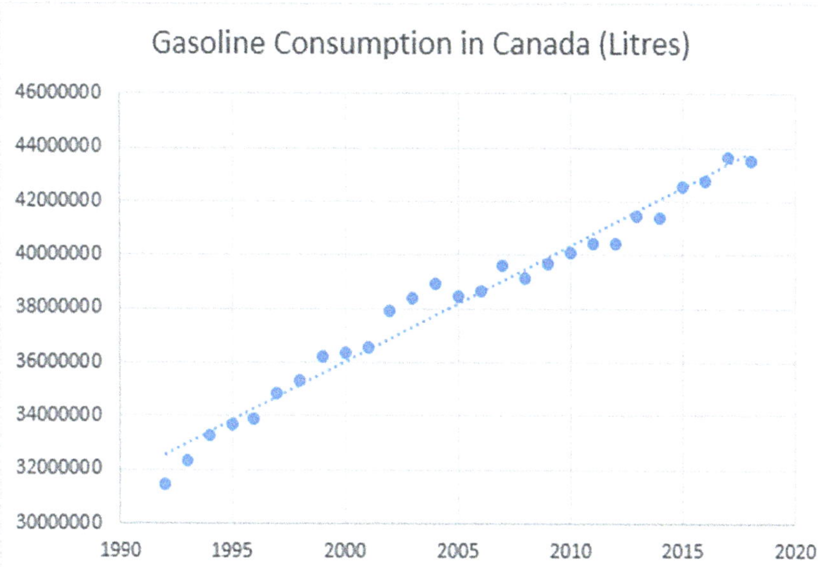

The first chart shows wildly fluctuating price of gasoline. The second shows gasoline sales. The sales chart shows no relation at all to price. Consumption did not go down during the high price peaks in 2011 – 2015, and it did not go up when price came crashing down. The overall rate just kept going up with rise in population and GDP. If there is no sensitivity to the "real" price, why would there be sensitivity to a fake increase in price, as proposed in the carbon tax plan?

It is still obvious that eventually, at a certain exorbitant level, price sensitivity would have to kick in. However, it is not likely to kick in at the rates of carbon tax currently being proposed.

Canada is a significant producer of oil and recent history shows almost comical bungling of oil-based projects. The heartland of extraction is the Alberta oil sands, which produce bitumen. Alberta is a landlocked province and as such has troubles getting its product to market. In 2013 a project called "Energy East" was proposed whereby mostly existing pipelines would be used to send bitumen east to refineries in Quebec and the maritime provinces. General opposition to pipelines, especially in Quebec, resulted in cancellation of the project by Transport Canada in 2017. The Quebec Premier Francois Legault was quoted as saying "We don't want your dirty oil". In the meantime, his province continues to get oil from Saudi Arabia instead. It goes without saying that oil shipped by trans-ocean tankers burn many times more gasoline in transport than a pipeline ever would.

Irving Oil, the main sponsor of the Energy East project, found an alternative. Instead of using existing pipelines, they will now send it through British Columbia to the West Coast, then use foreign tankers to go South down to the Panama Canal, through the canal to the Atlantic, and back up North to the refineries on the Canadian East Coast. The bitumen will now travel 11,771 kms, driven entirely by the burning of fossil fuels, instead of the 4600 km trip by pipeline. This solution met with no opposition from anyone.

In general, the mere hint of a new pipeline causes a rush of protests from many avenues, including the general public and environmental groups. Data shows that more and more oil is being moved by rail instead, and rail accidents involving spills and fires are much more frequent than spills from pipelines. The town of Lac-Megantic in Quebec was virtually destroyed by a derailed train carrying crude oil in 2013, causing the death of 47 residents.

Another pipeline project, the Coastal GasLink in British Columbia, was set to bring natural gas to a coastal plant where it would be liquefied and sent to customers in Asia. Over the long-term approvals were obtained from Indigenous peoples for the line to pass through their lands, but certain hereditary chiefs staged protests and blockades. They gained much sympathy from some quarters because, once again, pipelines were seen as contrary to climate change goals. It was not seen as relevant that the gas was destined for customers in Asia who would use it to replace coal-fired power plants, thereby producing a net reduction in GHG emissions.

Many groups created other blockades across the country in sympathy with the hereditary chiefs, causing considerable disruption of rail cargo transport. Eventually these came down. At time of writing, construction of the pipeline continues, but still without unanimous approvals. More trouble is likely yet to come.

One more story. Believe it or not, Canada is a net carbon sink! All of the country's emissions are more than offset by carbon absorption by plentiful forests and peatland. Some politicians have expressed dismay that Canada gets no credit for this in international monitoring of emissions. The real credit is derived from a relatively sparse population living mostly in cities, leaving plenty of wide-open spaces for forests to grow. Meanwhile, on a per capita basis Canada is actually the one of the world's biggest energy hog, on a par with the United States.

I relate all of these Canadian anecdotes because they represent a microcosm of climate change "strategy" throughout much of the world. In countries with functioning democracies, politicians have the usual short-term goals of getting elected and staying in power. At the same time, they strive to meet voters' inherently opposing views of strong climate change action without much personal sacrifice. This means signing up to international agreements but then doing little to bring down emissions. The public and the media have a love-hate relationship with fossil fuels, and a pure hate relationship with pipelines. Meanwhile pipelines are certainly the form of fuel transport that produces the lowest emissions and the lowest risk of leakage, and neither their existence nor non-existence has any real effect on fuel demand.

In countries without functioning democracies, the situation is of course bleaker. Decisions are made for personal or nationalistic interests, leaving the health of the planet in a decidedly secondary position.

THE INTERNATIONAL AGREEMENTS

In 1994 the United Nations Framework Convention on Climate Change (UNFCCC) was born. At that early stage of discussions, it has to be considered a moderate success in the sense that it formally recognized the issue and got most important emitting countries to become member states. It was modeled after the successful Montreal Protocol of 1988, which was responsible for setting standards to greatly reduce emissions that were destroying the ozone layer in the upper atmosphere.

The world has been fiddling with global climate change action plans ever since. The first agreement was the Kyoto Protocol in 1997. Although nearly 200 countries participated, the Protocol gave special considerations for underdeveloped countries that were recognized as lacking resources to commit to large binding reductions. In the end, 37 countries had binding commitments, and most met them by the target year 2012. Nine countries had to fund emissions reductions in other countries to meet their own targets, and Canada simply withdrew when it became obvious that they were not going to make it. Canada's commitment was 6% below 1990 emissions level by 2012, but in 2009 its emissions were already 17% above 1990 level. The United States signed the agreement but internally it was not ratified by the US Senate and therefore The US can also be considered withdrawn from the Protocol. The prime stated justification in the US Congress was that China, the world's biggest emitter, was let off the hook due to its status as a developing country.

There were updates to Kyoto targets and commitments, but it was clear that not enough progress was being made. In 2009 a new Accord was drawn up in Copenhagen. One hundred and fifteen world leaders met but they basically failed. There was a general agreement to hold temperature increase to 2°C, but no country would make any binding commitments. This was a first foray into international waters for the US Obama administration, but it produced nothing but the need to "keep on talking".

Today we have the Paris Accord of 2015. Under intense pressure to act from all quarters, an agreement was reached with 197 signatory nations. Not a bad achievement. The agreement allows each nation to set their targets, measure progress, and report back at regular intervals. There are no binding penalties for missing targets except the weight of international condemnation. Taken together, it was hoped that if all commitments are met, global temperature increase would be limited to less than 2°C. We will conclude on this after looking at some numbers.

As of 2017, 125 nations have internally ratified their commitments, and more are on the way. In 2020 the United States withdrew its support under the Trump Administration, but this has changed dramatically under the Biden Administration, which has promised targets even more aggressive than those in the Paris Accord.

Below are 2 charts, presented separately for clarity because of the different scales. They show historical emissions for various countries and also the commitments made for emissions reductions. Mostly countries have used the year 2030 as a common time for setting emissions targets, and the European Union (EU) also has a target for 2050. The vertical scale is Millions of Metric Tonnes of CO_2 equivalent, meaning that emissions from GHG's other than CO_2 have been converted to CO2 equivalency for accounting purposes.

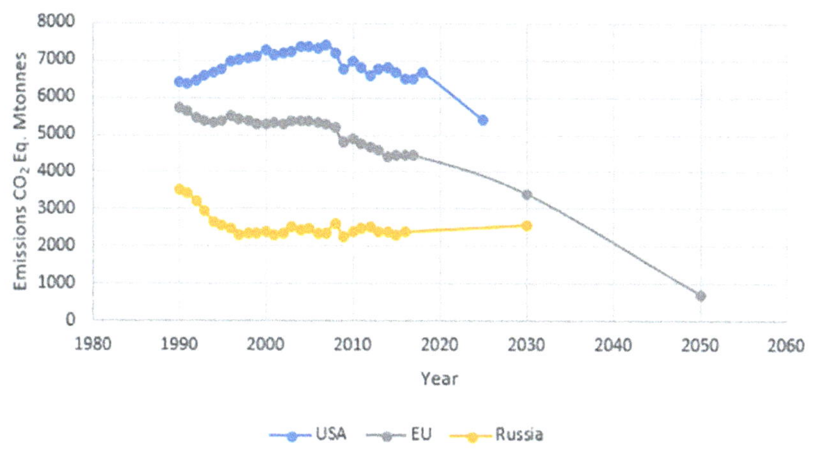

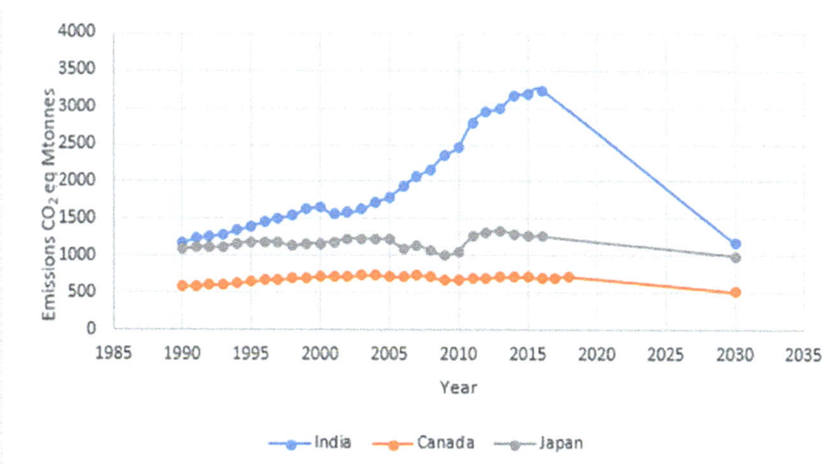

USA has shown a marked drop in emissions lately. It is mostly attributed to hydraulic fracking in the USA, which has led to a wealth of natural gas. The increased local availability of natural gas has led to a decline in the use of coal. The promised future reductions look aggressive, especially considering substantial opposition to cleaner energy among the political elite.

The EU has been something of a star in the international community. Led by Germany and the UK, there has been a consistent ongoing commitment to reduced energy intensity, increased efficiencies, and investments in renewable energy sources. The EU appears to be on track to meet commitments.

Russia's commitments, like many other nations, are with respect to a base year of 1990. As it happens, large reductions have already been made and therefore Russia has no real commitments to significant future reductions.

Canada and Japan have similar aggressive goals. I am not sure about Japan, but I see it as unlikely that Canada will reach its targets.

India is an important country because of its large population and its status as a developing country. As discussed in Chapter 4, the world would be in trouble if India went the way of China, where economic growth was accompanied by huge emissions. However, India has made the most aggressive commitments, and lately solar power is developing there at a tremendous rate. It still seems unlikely that they can meet these aggressive targets. Based on analysis in Chapter 5, there is a limit to how much difference solar power can make.

Now let us look at China.

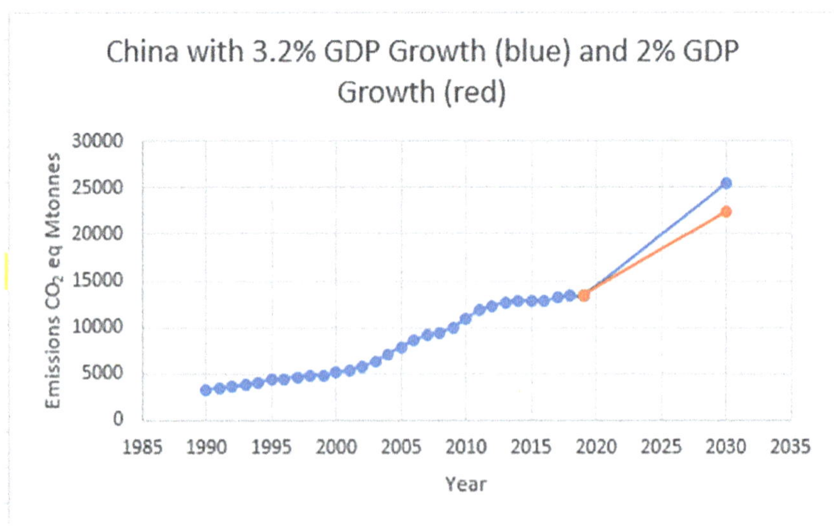

China's commitment as part of the Paris Accord was expressed as 60-65% reduction compared to 2005 emissions per unit of GDP by year 2030. That sounds like a lot, but the graph above shows what happens when you run the numbers. No-one can be sure what GDP will be in 2030, so I made two assumptions: GDP in China continues to increase by a current annual rate of 3.2%, or it increases annually by only 2%. The result shows that China's commitment is actually a massive increase in emissions.

In 2020 The Chinese President Xi Jinping made further commitments: China's emissions would level off by 2030, and the country would be carbon neutral by 2060. The first one does not seem too promising. Even my graph above would constitute leveling off in 2030.

The second promise of carbon neutrality has drawn much attention, and the reaction has been overwhelmingly positive. It is seen as putting more pressure on the USA to respond. I am not very sure myself. 2060 is 40 years away, and carbon neutrality does not mean zero emissions. It means that the amount of carbon released will be offset by carbon removed. Removal could mean paying another country for carbon offsets in an international trading agreement.

It should still be noted that China is indeed investing heavily in clean energy, and historically when the Chinese take on a large public project, they tend to succeed.

There is something else to consider about China. As everyone knows, a great deal of the world's manufacturing has shifted to China in the last 40 years. Chinese industry bears the burden of the energy costs of manufacturing so many goods. When other countries import those goods, are they not also responsible for the emissions used to make them? The USA outsources more and more to China and lives off of the corporate profits that results from cheap labor. They have a trade deficit with China of $350 Billion. If instead those goods were made in USA, American emissions would certainly be much higher. This illustrates how "we are all in this together".

The Paris Accord settles for the relative simplicity of accounting for only emissions created by each individual country, without the complexity of calculating emissions imports and exports. Such an accounting system does exist for some products, but I will not delve into that here.

A final graph below shows emissions for the rest of the world, meaning all of the countries not included in the discussions above.

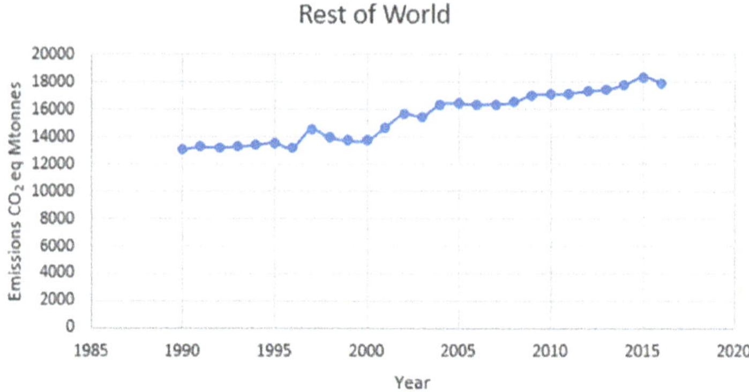

Here I have not extracted future targets from the many individual countries. I show this only to reveal that the historical upward trend continues. The magnitude represents around half of the world total. Most countries have chosen targets that are relative to "Business as Usual" (BAU). This means they project their emissions in 2030 based on recent history and commit to a percentage reduction of that number. Therefore, for example, a commitment to 25% reduction on BAU means somewhat less than 25% of current emissions.

Reading through the numbers, I see roughly an average commitment of about 25% of BAU, and I will assume this is equal to about 20% of current emissions by 2030. I use 2016 as the base year for this calculation. The Rest of World Chart would then look like this.

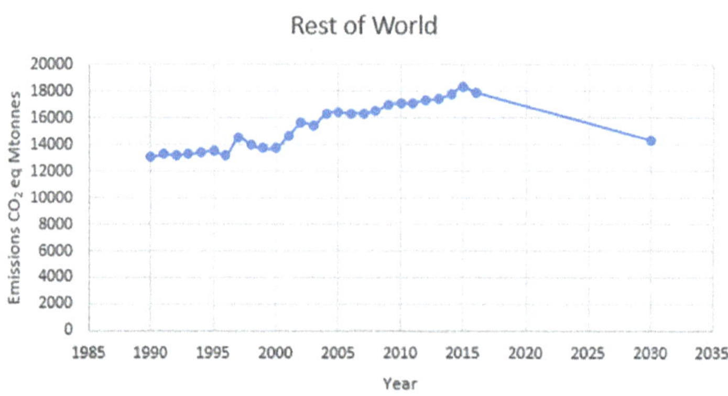

We see that the world will get back to emission levels seen in around years 1995 – 2000. This does not seem like much of an achievement since we know emissions in those years have aided in increasing today's global temperature.

We now could add up our gains and see where we are in 2030. The answer is: -1534 Mtonnes, an increase! This is because of China's "commitment" to increase emissions dramatically, and this factor dominates over all savings from the rest of the world. If we instead assume China's emissions do not increase at all, then the savings are 8041 Mtonnes, equivalent to .0043°C in temperature saving per year compared to staying steady at 2016 emissions level. If countries stay there until year 2100, the net savings in temperature increase will be .3°C. Recall that the target reduction should be around 2°C in order to avoid the catastrophic "Business as Usual" increase of 4°C that corresponds to RPC8.5 scenario.

Clearly, we need much more dramatic solutions than those currently planned in the Paris Accord. Even the .3°C saving is at risk, considering that China's emissions will not level off in 2020 as assumed there, and also considering extreme resistance to any radical changes in approach in the USA political regime.

Let us keep in mind that the targeted commitments might be considered aggressive. Reducing emissions by 20 to 30% while population and GDP are still growing is difficult, especially with the tools available. We saw in Chapter 5 that solar and wind cannot be expected to provide more than 40% of electricity production, and that produces only .24°C temperature reduction. Contributions from electric vehicles are almost negligible. The only really big hitter we have found so far is continuous increase in energy efficiency and reduction of power consumption.

In spite of these targets which I call aggressive, we need more options to bring down our emissions. We will explore those in Chapter 7.

SOCIAL ACTIVISM

In September 2019, a series of global Climate Strikes took place. Some 6 to 7 million people participated, leaving schools and work to take to the streets demanding action on Climate Change from their leaders. This was not the first nor the last, but it was the largest ever. The movement has accelerated thanks to calls from a Swedish teenager named Greta Thurnburg who has come to be an inspiration to the movement.

It is hard to know what effect these events have. Is it possible that the extraordinary agreements reached through the Paris Accord among virtually all nations is a result of citizen activism? We may never know for sure, but it certainly cannot hurt.

I wonder, however, if Climate Change has taken center stage at the expense of even more immediate issues. In the past social activism has had some extraordinary successes on more focused issues.

In the 1960's and 70's, smog caused mostly by car exhaust was a plague on large cities, and people got fed up. The pollutants were not only unsightly but harmful to humans and animals. I recall visiting Paris in 1980 and I was shocked by the consistent bad smell in the air, clearly caused by car exhaust. Demonstrations and a lot of bad press led to the introduction of the catalytic converter, eventually to all cars. It was a 1950's invention but the lead in gasoline had to be removed before it could work properly. Today, in spite of there being roughly double the number of cars on the roads, the air in our cities is very much cleaner. Cars still produce CO_2, of course, but technically CO_2 is not a directly harmful pollutant, and it is colorless and odorless.

Under pressure from their citizens, many countries chose to ban commercial whaling in the late 1960's, and in the late 1980's there was an international ban. Large whales had been hunted to near extinction, but today whale populations are more abundant. The humpback whale was thought to have dwindled to a population of about 5,000 in the 1960's. Today the estimated number is 80,000.

In 1974 it was discovered that chlorofluorocarbons (CFC's) were destroying the ozone layer in the upper atmosphere. Activism played a role in acting. Today the Montreal Protocol of 1990 has 197 signatories, and compliance with elimination of CFC's is universal. The ozone layer is obediently returning.

In all these cases and many more, social activism has played a role. However, these were focused issues with straightforward solutions. I wonder if today we should turn our attention to sub-issues of climate change instead of the whole beast, which is exceedingly difficult to tame. As an example, the depletion of the Amazon rainforests is a participant in climate change in the sense that the trees that absorb CO_2 are removed. Destruction of the forests is to either generate lumber or clear land for farming. As a byproduct, species extinction is rampant because so many rely on their small local ecosystems to survive. I wonder what would happen if those 7 million protesters that marched for climate change action would instead have surrounded Brazilian embassies and consulates worldwide to demand action. I am guessing this might have a strong effect, turning international attention to one glaring and very harmful practice with a relatively straightforward solution. Instead, Brazil could claim that their contribution to the bigger issue of global climate change is negligible compared to highly industrialized nations. In fact, this is the argument the USA has been using since the Kyoto Protocol: no deal without the highest emitter, China. Basically, Climate Change is such a big issue that it lets everyone off the hook. No one country can really solve it, so we can blame it on everyone else. But if we drill down and focus on sub-issues, we can exert more pressure.

In fact, in September 2020 a coalition of 230 environmental groups and a few Brazilian agrobusiness companies sent an open letter to recently (2019) elected president Bolsonaro urging him to halt the surge in deforestation in the Amazon. This got some press, but not as much as millions of people paying unwelcome visits to Brazilian embassies.

A similar issue exists in the Canadian and Russian North. Boreal forests make up 55% of Canada's land area, and Canada is the 2nd largest country in the world. Overall, Boreal forests sequester some 208 Billion tons of carbon, and as they grow, they sequester more. In Canada, much of it is on Federal land and is protected, but still there is logging and mineral extraction that disrupts the balance in a minority of the area.

In Russia, a majority of forested lands have been destroyed by logging and industry.

There is one sub-issue that would require international cooperation, but perhaps could be more easily tamed than Climate Change as a whole. Fish stocks have been depleted so badly that the UN has estimated that one third of fish species have been overfished. Overfishing means that the remaining population may be too sparse to replenish itself, and/or ecosystem shifts may occur due to the scarcity of the species. Oceans are primarily international waters and as such are difficult to police, even if we had laws to control fishing. However, if we dramatically reduced whaling, why not fishing also? It would again require international cooperation like with whaling and ozone depletion, but perhaps this is a task that could be achieved. Like Climate Change, it is actually in everyone's best interests. How does this relate to climate change? Well, not directly, but in the future the loss of marine life as a food supply would put even more pressure on farming, which is expected to see stress from droughts and freshwater shortages caused by Climate Change. Climate Change is not a monolithic problem. Everything in the Earth's ecosystems is connected.

The next issue is plastics in the ocean. There are misconceptions about the origins of all of this plastic. Most people in "wealthy" nations believe that they are contributing to change by using paper straws and re-usable water bottles. In fact, in those countries plastic waste goes either into recycling or landfill and never reaches the ocean. A quick Google search reveals that 90% of all plastics in the oceans originates from 10 rivers in Asia and Africa. These rivers pass through regions that are relatively impoverished and have used insufficient methods to dispose of their garbage. Typically, garbage dumps are close to the rivers themselves where the highest population density exists. Seasonal flooding causes overflow of garbage into the rivers, and hence into the oceans. I am not really sure how to tackle this, but perhaps wealthy nations should be aiding with managing garbage in these regions. This would certainly have more positive effect than banning single use plastic in areas where it can be readily recycled.

There is one issue where a sub-issue of Climate Change has already taken hold amongst activists. I touched on it briefly in my Canadian anecdotes at the beginning of this chapter. Many new fossil fuel projects, and especially pipelines, are met with demonstrations, resistance, and public condemnation. Unfortunately, I must hang my hat on the unpopular side of this issue. I have tried to demonstrate with facts in Chapters 4 and 5 that we are not ready now or even in the near future to abandon fossil fuels. The alternatives are simply not ready, short of dragging down economies and creating shortages in critical necessities, such as food supply. We need to treat fossil fuels as a precious resource until we can catch up with bringing other alternatives on board. It must be used as sparingly as possible, not only to limit climate effects but also to avoid prematurely creating critical shortages of fuels needed to sustain us.

There is a public perception that we can just "switch to renewables" now, and only the oil lobby is stopping us from doing it. It is certainly true that oil companies and lobbyists have campaigned with disinformation about climate change, especially in the 1980's and 1990's. However, that by itself does not mean that ready alternatives exist. As with many large issues in the public domain, there is a tendency to create a pure good and evil narrative. Fossil fuels and the oil companies are the evil, and renewable energy sources are the good. We need to drill down and look at everything with a bit more subtlety. There are no demons and also no saviors.

COMMUNICATION

A quick survey of people I know shows, as I expected, that awareness of details contained in Canada's climate action plan is extremely limited. A few have heard of carbon pricing; none have heard of any other element of the plan. A few months ago, this would have included me, even though I read through many national and international news services daily.

The Canadian Government's web site provides good, detailed information about what is being done at government level. A summary of Canada's plan can also be obtained by going through the Paris Accord documents. Therefore, information is available, but not well known or understood.

I believe a better communication strategy is needed. So does the United Nations. In July 2019 they issued an article with the following title: "Governments Agree to Strengthen Climate Education, Awareness and Public Engagement". Perfect. Reading through the article, it seems that some few nations like Chile have aggressive participatory education plans for their citizens. I am not sure though that the message is getting though globally.

What I envision is the leader of each country taking an active role in explaining the details of what is happening, and also what is not happening. A dialogue needs to be established with citizens in order to get their buy-in.

The exact mechanisms for such a dialogue are not clear to me, but the media must play a much stronger role than in the past. Reporting what politicians say and then getting stale commentary from guest speakers is not the way. Politicians need to be held to account, and that means doing some real investigative journalism. Regarding Canada's climate action plan discussed above, I found it on the government web site, and it is also available in summary form by going to the Paris Accord website. However, in an admittedly incomplete review I could not find any details in major Canadian media services, except for a couple of recent conservative opinion pieces claiming that renewable energy is too expensive. I did find many articles about carbon pricing, and even those were universally limited to the political implications of doing it or not doing it. No wonder that the public is uninformed and disengaged.

Wondering if this is a phenomenon unique to Canada, I went searching for what is happening in Germany. Angela's Merkel's Ruling conservative alliance, in coalition with the Social Democratic Party, put the Climate Action Law into effect in December 2019. It is important to note that this is a law, not a guiding policy document. A similar law was put into effect in the UK in 2009 and it has yielded impressive results. The German law details allowable emissions from all contributing portions of the economy, and it requires the government to report back on progress every 2 years. This law did not come about without opposition from within Germany. There were certainly complaints about the costs. Energy costs in Germany have already increased greatly as a result of the huge investments that were required in energy infrastructure. However, student activism was one of the drivers to push the law forward. A combination of a focused political regime and motivated citizens has led to one of the best performances in the world so far on emissions reduction, and the new law will probably ensure continued further progress.

I located a global survey done by Reuters (Nic Newman with Richard Fletcher, 2020) on the subject of Climate Change. Below is one of the results.

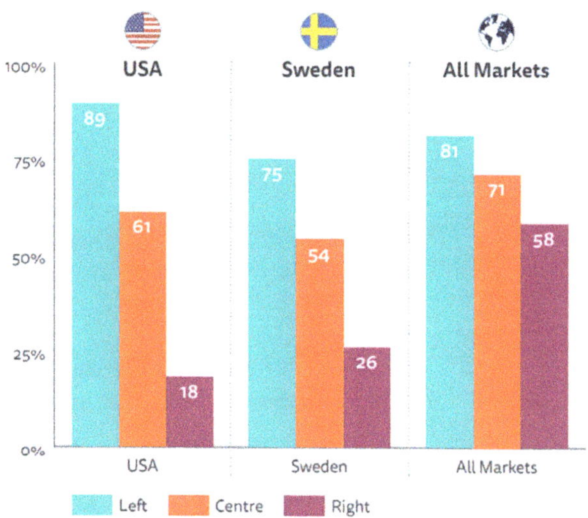

Not surprisingly, political leanings have a significant effect on views about Climate Change, especially in USA.

Below is a look at all countries surveyed.

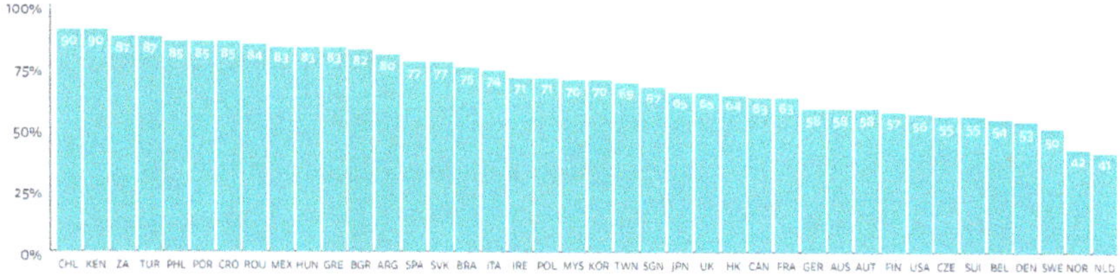

From this survey, it could be concluded that Germany and USA have similar views on the seriousness of Climate Change. Percent of the population that see it as very serious are 58 in Germany and 56 in USA. Therefore, we might conclude that the difference between Germany's strong actions and USA's weak actions lies not in the wishes of the populations, but in the contrasting wills of the political elites.

CONCLUSIONS FROM THIS CHAPTER

Political will for change varies greatly from country to country. In many cases such as Canada and USA there has been little more than lip service, while in the European Union there are commitments that have already been demonstrated.

Populations are universally aligned with a will to reduce emissions, but it is not clear that important personal sacrifices are expected.

China's emissions dominate the world and are projected to continue to grow.

The Paris Accord is a good achievement in the sense that it has brought everyone to the same table and have forced commitments country-by-country. On the other hand, the sum total of those commitments is insufficient to meet climate goals.

Social activism can have positive effects, but it is suggested that activists drill down to attack sub-issues directly. Climate change as a whole is a difficult target because it is too broad to have meaning to decisions makers.

The demonizing of fossil fuels and oil companies is not productive. These fuels are needed in the short and medium terms because the potential of renewable energy is insufficient to meet the demand or even to meet climate goals.

Greatly improved communication between politicians and citizens regarding national Climate Change plans is required to engage them in the battle to come. The media should take a stronger role in this process by assuming the bilateral roles of improved exposure of facts to the population and holding politicians to account.

CHAPTER 7

PROPOSALS FOR CHANGE

I WILL START THIS CHAPTER by explaining one by one the individual sources of emissions reduction that are available, including a re-visit of those already explored in Chapter 5. The reader will likely find that some of these items have not been well publicized, and I believe they need to have some light shed upon them.

As is evident from previous chapters, I will also insert opinions.

The strategies will be presented in roughly chronological order. Here I do not mean in the order of starting point, but the order of when benefits start to be realized. The starting points are all the same: right now.

Following the itemized list, I will attempt to present a comprehensive plan for the future, based upon everything I have learned on this journey.

ITEM 1: AGRICULTURE, FORESTRY AND OTHER LAND USE (AFOLU)

This is a broad category covering many sources and sinks of GHG's that will be described below. In summation, it covers an estimated 24% of all emissions, which is roughly equivalent to all electricity production. It deserves a great deal of attention.

Farming Practices

At the beginning of this journey, I must admit to having been almost entirely devoid of practical knowledge about farming and its effect on GHG's. I needed to find a good source of general information as a starting point, and I found one on the Government of Canada's website in a section called "Greenhouse Gases and Agriculture". It is a remarkably clear and detailed summary. I will steal some of the material for which I am thankful.

The figure below is a summary of the emission sources and sinks.

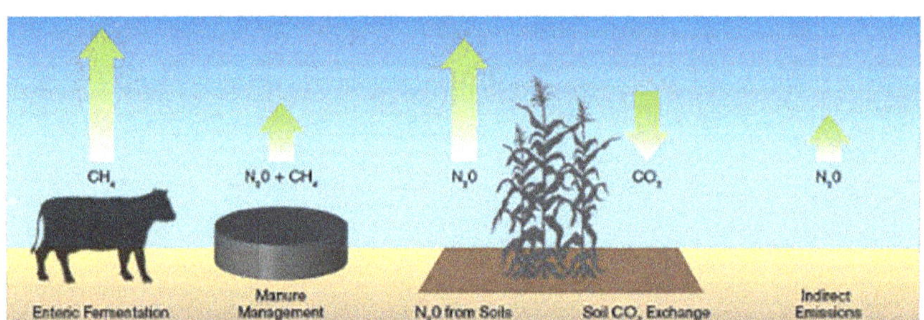

CO_2 is absorbed by crops as they grow through the normal photosynthesis process. The soil itself is also a carbon depository since all organic matter contains carbon. Plant litter, which is the portions of the plant remaining after harvest, also store carbon. On the other hand, the decay of plant matter releases CO_2 into the atmosphere. The processes used by farmers determines whether the farm will be a net emitter or a net sink for CO_2.

Farm practices that release more CO_2 into the atmosphere are:

Frequent tilling of the soil that exposes organic matter for faster decay

Leaving unproductive lands unused, so that no new carbon is added

Using excessive amounts of fertilizer to keep the land producing

Repeat plantation of the same crops onto the same land

Conversely, then, positive practices would be:

Reduced tilling

Crop Rotation

Replanting poor yield croplands with shrubs and trees to enhance carbon capture

Using animal manure instead of, or supplementary to, fertilizers

Surprisingly, these practices are expected to result in lower costs to the farmer and healthier soil.

Methane is released by farm animals as part of the digestion process. It is an indication that the digestion process is incomplete. Methane is also released by manure if it is stored in such a way that its decay is anaerobic. The solution to the second problem is of course the constant recycling of manure as fertilizer so that it does not need to be stockpiled or disposed of in dedicated dumps. The first problem is trickier, but scientists are in fact working on animal diet and other strategies to reduce methane emissions. One idea that is not circulating is to greatly reduce the number of cows for meat production, in favor of other animals or partial transition to vegetarian diet. Cows are by far the largest producer of methane among farm animals.

In addition to the methane emissions which are harmful, cattle require vast amounts of land for growing feed. These lands are often deforested lands that are converted to single crop production, which is doubly harmful.

Here is a graphic showing estimates of the world methane budget (Saunois, 2020). The ranges provided are indicators of uncertainty.

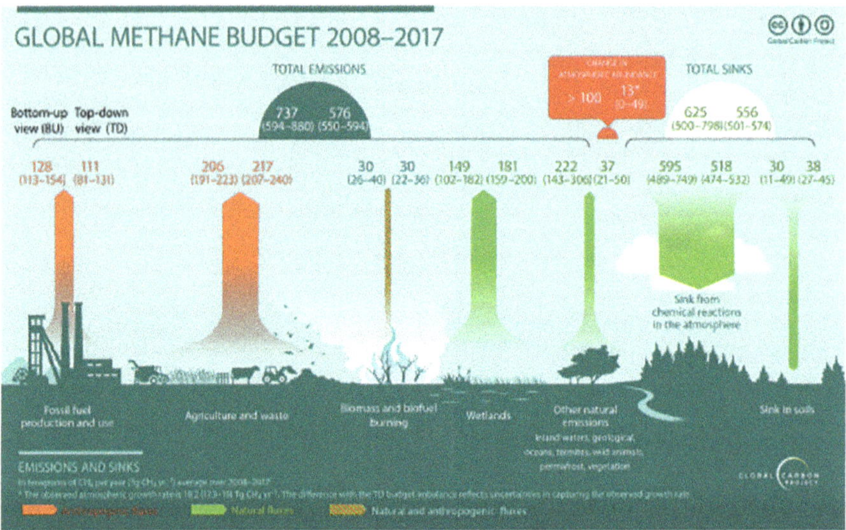

Agriculture is the main source of anthropogenic emissions, but fossil fuel extraction is also an important source, described as item 2 below.

Nitrous Oxide is produced naturally by soil in contact with oxygen, but it is greatly accelerated when fertilizers are added. An example of "careless" farming would be to add large amounts of fertilizer only once a year so that repeat application will not be needed. This leaves yearlong exposure to a large production of N_2O. A smarter way is to add only in small quantities, and only at the time when it is needed the most. Manure is also a better fertilizer in the sense that it causes lower concentrations of N_2O to be produced than chemical fertilizers. Finally, reduction in quantities of chemical fertilizers will have the added benefit of reduced amount of energy consumed in producing it.

Improving farming practices might seem like the simplest of all improvements to make, considering that the technologies are well understood and available. Eighty percent of the signatories to the Paris Accord have improved farming practices in their mitigation plans. I have reviewed some of the detailed plans and they all include the main strategies outlined above, such as crop rotation, manure management, and more focused application of fertilizers. However, government wish lists and delivering on promises can be different. Regulating farming practices might be required. Farmers need to get on board, and a good system of monitoring and reporting emissions needs to be maintained, such as the ones listed on the Canadian Government website.

- small chambers placed on soils or large chambers housing cows
- instrumented towers downwind of fields or instrumented aircraft flying over farming regions
- methods that require patient analysis of carbon change in soils over tens of years
- measurements of carbon dioxide (CO_2) in air, several times a second
- analysis of air in tubes buried in the soil, or from tubes hung high in the air on balloons

Trees and Forestry

Deforestation continues worldwide, although at a slower rate that in the 1900's. The primary reasons are logging for harvesting wood and clearing land for agriculture. In the accounting of GHG emissions, trees that are cut down "emit" the equivalent CO_2 that they would have absorbed from the atmosphere if they had been left to grow. In managed forestry new trees may be planted following harvest but they absorb less CO_2, in proportion to their size.

In this category fires that release CO_2 into the atmosphere are also included when they are deliberately set by humans. Fires are used as a tool to clear land prior to planting and to avoid some of the labor associated with cutting trees down.

The current rate of CO_2 processing by forests is around 4 Gtonnes per year. If this rate stayed the same from now until 2100, then that would become 320 Gtonnes, equivalent to .17°C in avoidance of temperature escalation.

The idea of tree planting, or "reforestation", has taken hold in public discourse, but there is not much action so far. In fact, the world is slowly losing more forest area, although at an increasingly lower rate. In 2019 Canadian Prime Minister Trudeau famously promised to plant 2 billion new trees in the following decade. There is no action so far but maybe he will catch up.

There are a few obvious obstacles to reforestation. Suitable land must be found that can sustain forests including the required biodiversity, and other interests might well compete for such land.

Here is a world map from 2005 with forests identified, produced by the Food and Agricultural Organization of the United Nations.

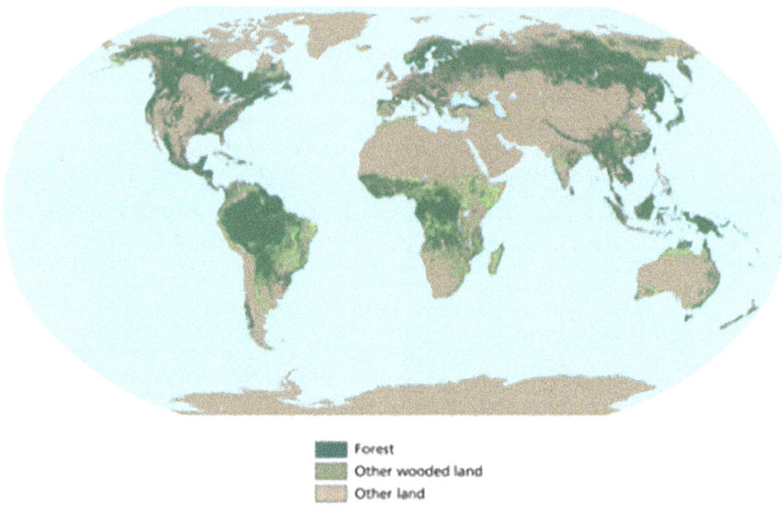

The map is dominated by the boreal forests in Canada and Russia and the South American and sub-Saharan rainforests. Now here is a map from Google showing desert regions.

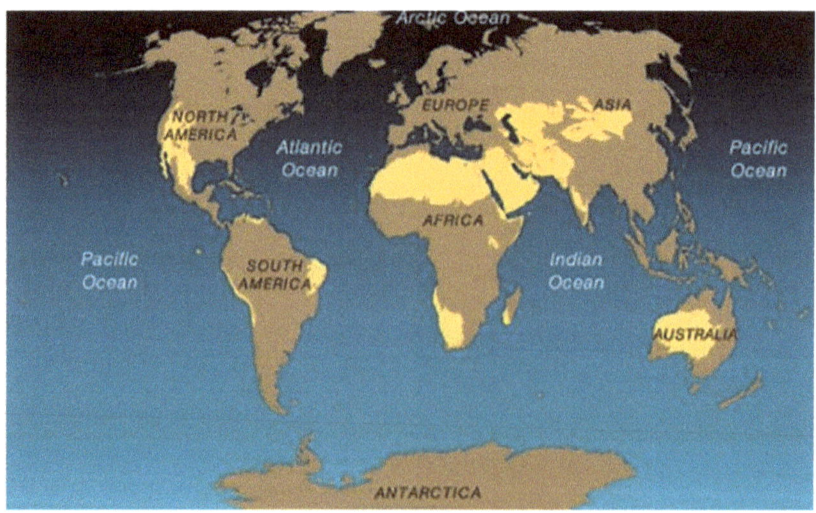

At this high level of observation, it looks like the world is already forested wherever it makes sense. Desert areas are certainly not candidates for forestation, and neither are heavily populated areas or the polar regions. Obviously, if we drill down to more local regions, there could be plenty of available land. However, it is unlikely that, for example, we could double the amount of forested area. Considering that all current forests provide only .17°C benefit over the next 80 years, then our best efforts will produce considerably less than that.

I would still favor taking on reforestation, with some benefits for Climate Change, and possibly major benefits for biodiversity. The most important action, however, is to immediately halt the destruction of existing forests that fall into the categories of tropical rainforest and northern boreal forest. This does not mean we end forestry. Harvesting of trees and replanting is not all bad in mid-latitudes. The new trees grow and absorb carbon, and the wood made from harvested trees is far superior to steel or cement for building construction, from GHG emissions point of view. However, we need to identify which lands can produce those trees in a sustainable way, and we need to stay out of pristine, natural forests.

Studies on the Amazon rain forest have shown that complete removal of all trees would disrupt the water cycle over a vast area that would likely revert to an arid environment. Considering that the prime reason for removing trees in the first place is for agriculture, it would be self-defeating if the daily rains are removed.

In Canada, much of the boreal forests are protected, but certainly not all. There is a hodgepodge of land ownership among federal and provincial governments and the private sector. The province of Quebec in particular is allowing more and more logging in Boreal forests. Since 1990, almost a quarter of the forest has been cut down, and there is no sign of restraint. Replanting here is not so simple. Because of the short growing season in the North, it takes an exceptionally long time for trees to develop. Growth in newly planted trees can be overwhelmed by shrubs which capture the light and nutrients from the soil. This is truly a region to be left alone. In Quebec's green plan announced in November 2020, no mention is made of these forests.

Costa Rica presents an astonishing success story in this category. A country that historically had 75% forest coverage slipped to 21% in 1987. A long succession of government actions reversed the trend. This included getting international loans for forestry management, some of which was literally handed over to the very people who were cutting down the trees to help them preserve the land instead. Today coverage is at about 60% and growing.

GHG Savings from AFOLU

Unlike the other sectors, projecting a specific benefit for things like improved farming and forestry practices is difficult. Here I rely instead on detailed calculations by many researchers and summed up in the IPCC reports (Smith P., 2014). For AFOLU the estimate of potential savings is 0.49 to 10.6 Gtonnes of CO_2 equivalent per year, starting in 2030. Note that Methane and Nitrous Oxide have been converted to CO_2 equivalent for accounting purposes. The difficulty in making accurate estimates is acknowledged in the report. For my purposes, I choose to use a low estimate of 5 Gtonnes due to low confidence level with so many factors and players involved. I will ramp up to this value from 2025 to 2030, then maintain it as a constant for the rest of the century.

ITEM 2: EMISSIONS FROM FOSSIL FUEL PRODUCTION

An important source of methane is leakage from fossil fuel extraction sites and industrial facilities that use these fuels. Natural gas has methane as its prime constituent. It is inevitable that in its extraction, transport, storage and use there will be some leakage. The US EPA estimates that 1.3% of natural gas drawn from underground reservoirs is eventually leaked out. This number could be as high as 3% when gas is obtained by fracking. Methane is also released from coal and crude oil extraction sites.

One mitigation strategy that has been used is to burn it off when there is a known source rich enough to ignite. This is called flaring. It still creates CO_2 but that is less damaging than methane in the atmosphere.

Leaking gas is estimated to be worth US \$30 Billion annually. Perhaps its pure economic value in conjunction with climate effects may convince oil and gas companies to start to take it seriously. Strategies to reduce these emissions include modernizing equipment to reduce leaks and monitoring extraction sites with methane detection equipment to find leakage sources. It seems that the technology to improve here is readily available. It may require government level regulation to make it happen.

Other emissions from the fossil fuel industry are caused by the energy needed to extract and transport fuel, and to convert crude oil to usable gasolines.

All of this amounts to a substantial 10% of total GHG emissions. There are already ongoing efforts to devise ways to reduce gas leakage at extraction sites and power plants. We could also assume that energy needed to extract fossil fuels will diminish with time as we ween ourselves off of them.

For my purposes, I will make the simple assumption that 2% of natural gas is leaking, and provisions will be put in place to reduce this to 1% by 2030. I will start a ramp up to this value starting in 2025. The 1% value will be maintained as a constant as a percentage of 2020 value, then starting in 2060 I will ramp it down to zero due to expected reduced usage of fossil fuels by then. I will also follow the same ramp-down formula for the other emissions associated with fossil fuel production and transport.

ITEM 3: CARBON CAPTURE AND SEQUESTRATION

One way to be able to continue burning fossil fuels without CO_2 emissions is to capture those emissions, either at the combustion source or from ambient air. There are many different processes that can be used. There are pre-combustion processes, where the fuel, gas, or pulverized coal, is separated into CO_2 and hydrogen H2, and then the H2 is used to generate power while the CO_2 is removed. Burning of H2 is clean, with water vapor as the byproduct. Then, there are post-combustion processes where the exhaust flues from the power plants are captured and processed to chemically remove the CO_2. This is advantageous because existing power plants can be adapted by adding a separate facility to process the exhaust gases from the main power plant.

All of these processes require energy inputs. In other words, more fuel needs to be burned than would otherwise be required to produce the same power output. In the 2005 IPCC report (Metz, 2005) on carbon capture the additional fuel needed is in the range of 24% to 40%, depending on the fuel and the process used. Here is a table from a later 2019 report (Haseli, 2019) which extracts data from different studies on the available processes.

Fuel Type	Process	Net Efficiency	Net Power (MW)
Coal (Bituminous)	Without carbon capture	44	758
	Precombustion	31.5	676
	Post combustion	34.8	666
	Oxy Combustion	35.4	532
	Oxy Combustion (Allam Cycle)	51	226
	Chemical looping combustion	44	115.5
Natural Gas	Without carbon capture	55.6	776
	Pre combustion	41.5	690
	Post combustion	47.4	662
	Oxy Combustion	44.7	440
	Oxy Combustion (Allam Cycle)	59	303
	Chemical looping combustion	52.2	364

To interpret this table, I refer to the line "Post-Combustion" with coal as the fuel. The efficiency of the plant is 34.8%, compared to 44% for a plant without carbon capture. Furthermore, the net power output, delivered to the grid, is reduced from 758 MW to 666 MW. In order to deliver the full 758 MW at the reduced efficiency level, a total of 44% more fuel needs to be added. This seems to be in line with the IPCC report.

Following CO_2 extraction, it needs to be compressed into a form that is transportable in sensible quantities, and then delivered to a site where it can be sequestered. The transport by truck or rail uses still more fuel, producing CO_2 which is not captured. This might be only a few more percentage points of the plant output.

A related technology is carbon capture from ambient air. Again, there are a few different processes being considered, but it evidently also requires energy input. Because of the extremely small concentration of CO_2 in ambient air, huge amounts of air need to be moved to capture it in significant quantities.

Sequestration presents its own problems. Ideal locations are depleted fossil fuel extraction sites where the CO_2 can be pumped down into the chambers or porous rock formations originally occupied by oil or natural gas. The site must then be capped to avoid leakage back out into the atmosphere. These ideal locations might well be located far from the power plants where the CO_2 was extracted. Storage on ocean floors has also been considered. If injected at great depths, the CO_2 will supposedly not leak into the atmosphere for many decades or centuries. I would still worry about ocean acidification caused by CO2 uptake in the water, and its effect on marine life existing at great depths.

The amount of CO_2 captured is reported in theoretical studies as around 90%. However, that is 90% of a larger number because of the increased fuel input to the cycle. That means that emissions would not be 10%, but 12% to 15% compared with power plants with no carbon capture.

In a real-world case, a carbon capture facility called Petra Nova went online in 2016. According to the Institute for Energy Economics and Financial Analysis (IEEFA), the plant only captures 65 to 70% of CO_2, and this does not consider the additional CO_2 produced by the natural gas turbine that powers the carbon capture module.

Perhaps these issues will get resolved. More efficient ways to extract CO_2 may be developed and become routine. However, I feel reluctant to take credit for this technology at this time. The idea of burning significant amounts of extra fuel to extract CO_2 feels wrong. There are other damaging emissions from this extra fuel, especially from coal, which are harmful to humans and the environment. Furthermore, if we do this on a large scale, we will be further depleting the fossil fuel resources which, as I have claimed in previous chapters, are not infinite and are at risk of dangerous scarcity on the time scales that are relevant here.

For these reasons I do not plan to include carbon capture as an opportunity in my suggested plan at the end of this chapter.

ITEM 4: HYDROGEN

Hydrogen H is the lightest and most abundant element in the universe. It does not exist in its elemental state because it readily combines with a neighboring atom to form the molecule H_2, similar to other gases. In the presence of oxygen, it also typically does not exist because it is very volatile. Molecular hydrogen is an excellent fuel. When it burns, the only product is water vapor. However, it must be extracted from another source, compressed to have a useful amount in a small volume, and contained in an oxygen-free environment.

There are two principal methods to obtain H_2 as a fuel. One is to extract it from other liquid or gaseous fuels such as gasoline or natural gas. This requires energy input, as mentioned above in the section on carbon sequestration. Furthermore, the net amount of CO_2 produced is greater than the burning of the fossil fuel directly. This is clearly not a desirable option.

The second method is by electrolysis of water. This seems like an attractive option, because the input is water, and the output is simply more water being delivered back into the biosphere. However, it is important to realize that molecular hydrogen is really just an energy storage medium. There is always energy required to obtain it. In fact, more electrical energy needs to be input by electrolysis than is available from the hydrogen gas obtained. If the electricity comes from the burning of fossil fuels, then this option is nonsensical in the context of global warming.

Once the gas is obtained, it must be compressed because at ambient pressure there is not enough of it to be useful as a fuel. Typical compression ratios considered are between 200 and 800. Then, once it is compressed and stored in a pressure vessel of some sort, it needs to be transported to where it can be used, such as an industrial site or a vehicle filling station. Compression and transport all require even more energy.

All of this was evaluated in a study (Ulf Bossel, 2001) which produced the following table. Path A considers the compression of H_2 as a gas and delivered to end users, path B considers the chilling and liquefaction of H_2 for transport, and path C considers production of H_2 on site to provide fuel to vehicles.

	Energy cost in HHV of H_2	Factor	Path A gas	Path B liquid	Path C onsite
Production of H_2					
Electrolysis	43%	1.43	1.43	1.43	
Onsite production	65%	1.65			1.65
Packaging					
Compression 200 bar	8%	1.08	1.08		
Compression 800 bar	13%	1.13			
Liquefaction	40%	1.40		1.40	
Chemical hydrides	60%	1.60			
Distribution					
Road, 200 bar H_2, 100 km	6%	1.06	1.06		
Road, liquid H_2, 100 km	1%	1.01		1.01	
Pipeline, 1,000 km	10%	1.10			
Storage					
Liquid H_2, 10 days	guess: 5%	1.05		1.05	
Transfer					
200 bar to 200 bar	1%	1.01	1.01		1.01
Delivered to User					
Energy Input to HHV of H_2			1.65	2.12	1.66

For clarity, path C also uses electrolysis like path A, but the onsite entity is a vehicle filling station where the relatively small production quantities lead to some inefficiencies.

The conclusion is contained in the last line. It requires between 165% and 212% of the energy contained in the H_2 fuel to extract it using electrolysis, compress it into usable quantities and transport it to end users.

Why would we do this? There is only one sensible scenario. If the electricity being used to extract and compress H_2 comes from a "clean" source, then this is a way to get a mobile fuel without CO_2 emissions. In Australia, a new project called the Asian Renewable Energy Hub is currently in development. Part of this project will produce hydrogen by electrolysis of water using electricity generated by solar and wind power. Some of the Hydrogen is expected to be transported to Japan and Korea who will use it as part of their green energy transitions. In Germany, Angele Merkel has also announced plans to start building "Green Hydrogen" infrastructure. In Saudi Arabia, a new city is being built around a green Hydrogen project.

In car applications, hydrogen would consume much more energy than battery-powered Electric Vehicles (EV). Electricity coming from the grid and used to charge batteries is almost all used in motive power. Electric motors are typically more than 90% efficient. Conversely, Hydrogen requires more than 150% of its stored energy to produce, and if used in an internal combustion engine, it also produces waste heat. Improved efficiency is obtained in a hydrogen fuel cell vehicle, where Hydrogen generates electricity that drives the vehicle through the same type of efficient electric motor used in EV's.

One big advantage for hydrogen in the vehicle market is greatly reduced weight. In Chapter 5 we found that a EV with sensible range can require a battery weighing around 500 kgs. The hydrogen, storage tank and fuel cells that would replace such a battery would weigh only about 120 kgs.

So, if we had large quantities of emissions-free electricity, would we use EV's or Hydrogen powered vehicles? Hydrogen would use more of that clean energy, but EV's would be heavier and, as we saw in Chapter 5, they run out of usefulness for large vehicles. Perhaps we would use a mix of both types of vehicles depending on size and application.

If we can find clean electricity, hydrogen would definitely find a home in industrial applications that currently use coal or natural gas. This would be a bigger hitter for climate change than all the cars in the world. In the IPCC climate mitigation report (Fischedick M., 2014) Industry provides 15 Gtonnes of CO2 equivalent per year. Approximately 10 Gtonnes of that could be removed by the application of green hydrogen in burners and boilers that are used for materials processing and for generation of industrial heat and electricity.

In the plan, green hydrogen will only be used when excess capacity of clean electricity is available to produce it. It would be used to replace fossil fuels in industrial applications, large transport vehicles and a portion of the personal vehicle market.

ITEM 5: NUCLEAR - FISSION - URANIUM

Nuclear power has been with us since the 1950's. The technology was developed in the 1940's which at that time was dedicated to creating nuclear bombs. The process involves splitting heavy atoms into lighter ones, and in the process subatomic particles are released, which in turn create more collisions and more fission in what is called a chain reaction. Here is a graphic showing the reaction equation.

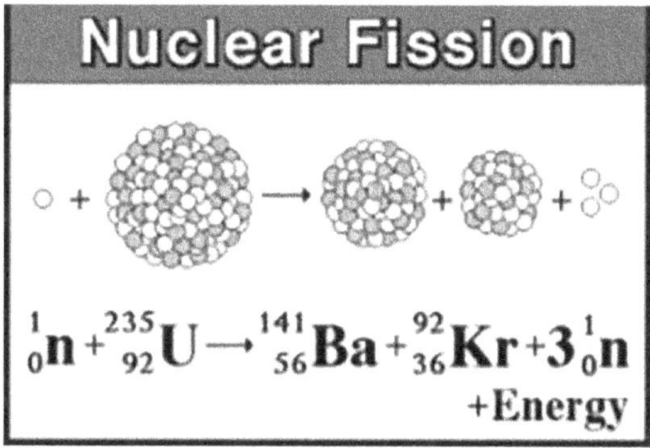

$$ _0^1 n + _{92}^{235}U \rightarrow _{56}^{141}Ba + _{36}^{92}Kr + 3_0^1 n + Energy $$

In nuclear power plants, the heat created by fission is carried away by water that surrounds the reaction chamber. The water becomes steam which drives a conventional steam turbine to generate electricity. In order to control the reactions, control rods are inserted into the chamber which serve to "catch" some of the escaping particles, preventing them from creating more fission reactions.

The element most amenable to fission is uranium, atomic number 92 and atomic mass 238 on the Periodic Table. The isotope U235 is used in nuclear power reactors. Uranium exists in the Earth's crust at an average concentration of 2.7 ppm. Most Uranium mines have 1% or greater concentrations.

In Europe and North America nuclear power is no longer seen as a viable alternative. On the other hand, many countries have ambitious plans to build new reactors, especially China. The key driver for nuclear power falling out of favor in some areas is the risk of accidents that could release dangerous radiation. There have been three such accidents.

In 1979 one of two reactors at Three Mile Island in Pennsylvania experienced a partial meltdown. It was caused by a series of mechanical failures, instrument malfunctions and human errors. A water pump that feeds the reactor broke down, leading to overpressure in the reactor piping. A release valve opened automatically to lower the pressure but failed to re-close to keep water in the core. Instruments told operators that the valve was closed, and they assumed there was water to contain the heat, but water flowed out of the open valve. No-one was hurt and the amount of radiation that escaped the plant was minimal.

In 1986 a much worst event occurred at Chernobyl in the Ukraine. The sequence of events was complex, but the root cause is known to be a combination of poor design and operator errors. There is an excellent HBO TV series describing the events. Twenty-eight of the on-site workers were killed, and another 106 received exposure high enough to cause severe radiation sickness. Most of the surrounding population did not receive any severe exposure. There has been continuous monitoring of populations for long term effects, but it seems that these have been less severe than anticipated. This is the only nuclear power event that caused fatalities, and the reactor design flaws do not exist today, neither in Ukraine or anywhere else in the world.

In 2011 the Tohoku earthquake was detected and caused a shutdown of Fukushima nuclear power plant in Japan. Back-up generators were used to maintain pumping of cooling water, but then the tsunami created by the earthquake struck, shutting down the diesel generators. Residual heat in the reactor cores was enough to cause meltdown of the containment structure. Radiation was released into the atmosphere and contaminated water into the Pacific Ocean. The surrounding areas were evacuated, and no deaths were recorded from this incident. Clean up and containment of contaminants continues to this day.

These events have led to improved design practices and safety protocols in the nuclear industry. However, in the eyes of much of the public and environmental groups the prospect of even worse disasters remains. Throughout the voluminous discussions on Climate Change, the nuclear option barely registers. Still, it is a fact that GHG's from nuclear power plants are a tiny fraction of those from fossil fuel plants.

From here I will use data from the World Nuclear Association, which is freely available on their website.

Let us look at the entire safety record of Nuclear power plants. To clarify, one reactor operating for 10 years or 10 reactors operating for one year would result in 10 reactor years on the chart below.

Cumulative Reactor Years of Operation

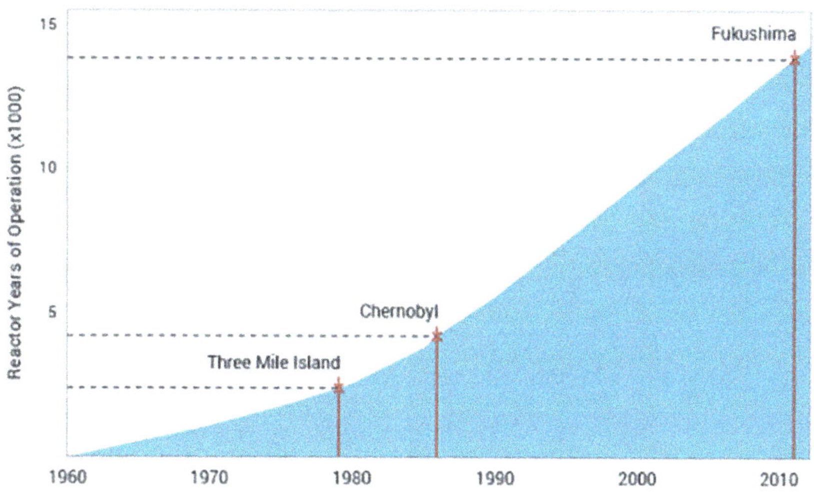

The world is currently approaching 17,000 reactor years of operation with the 3 incidents highlighted. This is actually a strong safety record. Considered another way, if there was only one reactor operating, it would have had only 3 incidents in 17,000 years to produce the same statistical probability of occurrence.

In my opinion, we need to consider that no technology is without risks and negative impacts. Some negative aspects can get exaggerated compared to others. Take for example airplane crashes. If a large airliner crashes it can kill hundreds of people, and that gets a lot of attention, as it should. Consequently, every major event results in an investigation that frequently leads to modifications of the rules for design and operation of aircraft. This closed loop process has led to a continuous decline in events over the decades, especially when expressed as a percent of passenger miles. In 2019, a total of 257 deaths occurred in plane crashes worldwide. Included is the Ethiopian Air crash of a Boeing 737 Max that killed 157 people. This received all the attention that it should; but consider that in the USA alone 38,800 deaths occurred in road accidents in 2019. That is equal to about one 9/11 event per month. This gets much less attention.

Nuclear power suffers from the same perception issue. Because a single event can affect a great number of people, it is perceived as being too dangerous. But in fact, its safety record is outstanding.

This leads me to suggest that nuclear power should remain on the agenda as a powerful opponent to Climate Change. It is a mature technology with all of the safety risks well understood. This does not mean zero risk, it means low, manageable risk.

With that in mind, let us look at the possibilities.

The known reserves of Uranium are at 6,142,600 tons. New deposits are known to be there but will require ever-increasing costs to extract because of lower concentrations as compared to existing sites. Another interesting source of uranium is the decommissioning of nuclear weapons following a 1987 disarmament treaty between USA and USSR.

Currently all the reactors in the world use 67,500 tons of Uranium each year. This means that the current known reserves can feed existing reactors for 91 years, or perhaps a little less because some is used for other purposes. Therefore, there is plenty available to increase usage of nuclear power substantially. Simplistically, if we tripled usage, we could supply them for 30 years, not including discovery of new uranium ore deposits. Nuclear energy currently provides 10% of electricity, so this could go to around 30% after the years of new reactor construction.

In fact, new ones are being built. Here is a summary of the plans from each country.

WORLD NUCLEAR ASSOCIATION

Country Profile(s)	NUCLEAR ELECTRICITY GENERATION		OPERABLE		UNDER CONSTRUCTION		PLANNED		PROPOSED		URANIUM REQUIRED
	TWh	% e	No.	MWe net	No.	MWe gross	No.	MWe gross	No.	MWe gross	tonnes U
Argentina	[illegible]	[illegible]	[illegible]	[illegible]	[illegible]	[illegible]	[illegible]	[illegible]	[illegible]	[illegible]	[illegible]
Armenia	[illegible]	[illegible]	[illegible]	[illegible]	[illegible]	[illegible]	[illegible]	[illegible]	[illegible]	[illegible]	[illegible]
Bangladesh	[illegible]	[illegible]	[illegible]	[illegible]	[illegible]	[illegible]	[illegible]	[illegible]	[illegible]	[illegible]	[illegible]
Belarus	[illegible]	[illegible]	[illegible]	[illegible]	[illegible]	[illegible]	[illegible]	[illegible]	[illegible]	[illegible]	[illegible]
Belgium	[illegible]	[illegible]	[illegible]	[illegible]	[illegible]	[illegible]	[illegible]	[illegible]	[illegible]	[illegible]	[illegible]
Brazil	[illegible]	[illegible]	[illegible]	[illegible]	[illegible]	[illegible]	[illegible]	[illegible]	[illegible]	[illegible]	[illegible]
Bulgaria	[illegible]	[illegible]	[illegible]	[illegible]	[illegible]	[illegible]	[illegible]	[illegible]	[illegible]	[illegible]	[illegible]
Canada	[illegible]	[illegible]	[illegible]	[illegible]	[illegible]	[illegible]	[illegible]	[illegible]	[illegible]	[illegible]	[illegible]
China	[illegible]	[illegible]	[illegible]	[illegible]	[illegible]	[illegible]	[illegible]	[illegible]	[illegible]	[illegible]	[illegible]
Czech Republic	[illegible]	[illegible]	[illegible]	[illegible]	[illegible]	[illegible]	[illegible]	[illegible]	[illegible]	[illegible]	[illegible]
Egypt	[illegible]	[illegible]	[illegible]	[illegible]	[illegible]	[illegible]	[illegible]	[illegible]	[illegible]	[illegible]	[illegible]
Finland	[illegible]	[illegible]	[illegible]	[illegible]	[illegible]	[illegible]	[illegible]	[illegible]	[illegible]	[illegible]	[illegible]
France	[illegible]	[illegible]	[illegible]	[illegible]	[illegible]	[illegible]	[illegible]	[illegible]	[illegible]	[illegible]	[illegible]
Germany	[illegible]	[illegible]	[illegible]	[illegible]	[illegible]	[illegible]	[illegible]	[illegible]	[illegible]	[illegible]	[illegible]
Hungary	[illegible]	[illegible]	[illegible]	[illegible]	[illegible]	[illegible]	[illegible]	[illegible]	[illegible]	[illegible]	[illegible]
India	[illegible]	[illegible]	[illegible]	[illegible]	[illegible]	[illegible]	[illegible]	[illegible]	[illegible]	[illegible]	[illegible]
Iran	[illegible]	[illegible]	[illegible]	[illegible]	[illegible]	[illegible]	[illegible]	[illegible]	[illegible]	[illegible]	[illegible]
Japan	[illegible]	[illegible]	[illegible]	[illegible]	[illegible]	[illegible]	[illegible]	[illegible]	[illegible]	[illegible]	[illegible]
Jordan	[illegible]	[illegible]	[illegible]	[illegible]	[illegible]	[illegible]	[illegible]	[illegible]	[illegible]	[illegible]	[illegible]
Kazakhstan	[illegible]	[illegible]	[illegible]	[illegible]	[illegible]	[illegible]	[illegible]	[illegible]	[illegible]	[illegible]	[illegible]
Korea RO (South)	[illegible]	[illegible]	[illegible]	[illegible]	[illegible]	[illegible]	[illegible]	[illegible]	[illegible]	[illegible]	[illegible]
Lithuania	[illegible]	[illegible]	[illegible]	[illegible]	[illegible]	[illegible]	[illegible]	[illegible]	[illegible]	[illegible]	[illegible]
Mexico	[illegible]	[illegible]	[illegible]	[illegible]	[illegible]	[illegible]	[illegible]	[illegible]	[illegible]	[illegible]	[illegible]
Netherlands	[illegible]	[illegible]	[illegible]	[illegible]	[illegible]	[illegible]	[illegible]	[illegible]	[illegible]	[illegible]	[illegible]
Pakistan	[illegible]	[illegible]	[illegible]	[illegible]	[illegible]	[illegible]	[illegible]	[illegible]	[illegible]	[illegible]	[illegible]
Poland	[illegible]	[illegible]	[illegible]	[illegible]	[illegible]	[illegible]	[illegible]	[illegible]	[illegible]	[illegible]	[illegible]
Romania	[illegible]	[illegible]	[illegible]	[illegible]	[illegible]	[illegible]	[illegible]	[illegible]	[illegible]	[illegible]	[illegible]
Russia	[illegible]	[illegible]	[illegible]	[illegible]	[illegible]	[illegible]	[illegible]	[illegible]	[illegible]	[illegible]	[illegible]
Saudi Arabia	[illegible]	[illegible]	[illegible]	[illegible]	[illegible]	[illegible]	[illegible]	[illegible]	[illegible]	[illegible]	[illegible]
Slovakia	[illegible]	[illegible]	[illegible]	[illegible]	[illegible]	[illegible]	[illegible]	[illegible]	[illegible]	[illegible]	[illegible]
Slovenia	[illegible]	[illegible]	[illegible]	[illegible]	[illegible]	[illegible]	[illegible]	[illegible]	[illegible]	[illegible]	[illegible]
South Africa	[illegible]	[illegible]	[illegible]	[illegible]	[illegible]	[illegible]	[illegible]	[illegible]	[illegible]	[illegible]	[illegible]
Spain	[illegible]	[illegible]	[illegible]	[illegible]	[illegible]	[illegible]	[illegible]	[illegible]	[illegible]	[illegible]	[illegible]
Sweden	[illegible]	[illegible]	[illegible]	[illegible]	[illegible]	[illegible]	[illegible]	[illegible]	[illegible]	[illegible]	[illegible]
Switzerland	[illegible]	[illegible]	[illegible]	[illegible]	[illegible]	[illegible]	[illegible]	[illegible]	[illegible]	[illegible]	[illegible]
Thailand	[illegible]	[illegible]	[illegible]	[illegible]	[illegible]	[illegible]	[illegible]	[illegible]	[illegible]	[illegible]	[illegible]
Turkey	[illegible]	[illegible]	[illegible]	[illegible]	[illegible]	[illegible]	[illegible]	[illegible]	[illegible]	[illegible]	[illegible]
Ukraine	[illegible]	[illegible]	[illegible]	[illegible]	[illegible]	[illegible]	[illegible]	[illegible]	[illegible]	[illegible]	[illegible]
UAE	[illegible]	[illegible]	[illegible]	[illegible]	[illegible]	[illegible]	[illegible]	[illegible]	[illegible]	[illegible]	[illegible]
United Kingdom	[illegible]	[illegible]	[illegible]	[illegible]	[illegible]	[illegible]	[illegible]	[illegible]	[illegible]	[illegible]	[illegible]
USA	[illegible]	[illegible]	[illegible]	[illegible]	[illegible]	[illegible]	[illegible]	[illegible]	[illegible]	[illegible]	[illegible]
Uzbekistan	[illegible]	[illegible]	[illegible]	[illegible]	[illegible]	[illegible]	[illegible]	[illegible]	[illegible]	[illegible]	[illegible]
WORLD	[illegible]	[illegible]	[illegible]	392,315	[illegible]	[illegible]	[illegible]	[illegible]	[illegible]	[illegible]	[illegible]
	TWh	% e	No.	MWe	No.	MWe	No.	MWe	No.	MWe	tonnes U
	NUCLEAR ELECTRICITY GENERATION		OPERABLE		UNDER CONSTRUCTION		PLANNED		PROPOSED		URANIUM REQUIRED

Looking at the line near the bottom specified as "World", we see 2020 energy delivery from reactors already in operation is at 392,315 MW, while a total of 461,119 MW capacity is either under construction or planned. It looks like we are already on our way.

In Most European countries and in North America, large efforts are being placed on wind and solar power, and almost none on nuclear power, presumably because of safety concerns. Climate Change mitigation plans in those countries could be substantially improved with new nuclear projects. We have seen in Chapter 5 that wind and solar will reach limits due to the intermittent nature of the energy they can deliver.

ITEM 6: NUCLEAR - FISSION - THORIUM

Thorium (Atomic Number 90, Atomic Mass 232) presents an alternative to Uranium for nuclear power. It is not fissile by itself, but it can be used to breed Uranium 233 which then becomes the fuel. Another fissile material must be used to generate the sub-atomic particles that will convert Thorium 232 to Uranium 233. Here is the reaction equation.

$$\overset{neutron}{n} + {}^{232}_{90}\mathrm{Th} \longrightarrow {}^{233}_{90}\mathrm{Th} \overset{\beta^-}{\longrightarrow} {}^{233}_{91}\mathrm{Pa} \overset{\beta^-}{\longrightarrow} \overset{fuel}{{}^{233}_{92}\mathrm{U}}$$

I will not describe here the details, but I list below the expected advantages of Thorium over Uranium.

1. Thorium is 3 to 4 times as plentiful in the Earth's crust as Uranium
2. Reactor byproducts cannot be used directly to make weapons grade fissile material.
3. The possibility of dangerous uncontrolled meltdown is greatly reduced.
4. Depleted reactor products are less radioactive and have much smaller half-lives than those produced by a U235 reactor.

With all of these advantages, one wonders why we have not gone this way already. In the USA, the early years of nuclear power were dominated by U235 reactors because they could produce plutonium for nuclear bombs, and also the process was better understood. Today there is a considerable amount of research needed to get Thorium reactors up and running, and it will be more expensive. With somewhat reduced interest in nuclear power in some parts of the world, Thorium power has repeatedly stalled, but it is not going away. In fact, India has made the bold claim that Thorium reactors will produce 30% of their electricity by 2050. China is also active in developing Thorium-based power.

If nuclear power has a strong resurgence under Climate Change pressures, Thorium could conceivably take over from Uranium as the century progresses, with no concerns over diminishing supply of material. A great deal of investment would be needed, both for the research efforts and for the construction of Thorium-based reactors, which are conceptually different from conventional Uranium-based ones.

ITEM 7: NUCLEAR – FUSION

While heavy elements like Uranium produce heat when they are split into lighter ones, lighter elements do so when they are fused into heavier ones. All elements up to and including iron in the periodic table are fused from lighter ones inside stars. This fusion process is what keeps stars burning. Elements heavier than iron are only produced by implosion of dying stars. It is fascinating to consider that the heavy elements we have on Earth like copper, silver, gold, and lead were all produced by stars that are already dead and gone.

In order to fuse elements, it is necessary to "cross the Coulomb barrier". The positive electrostatic forces that keep protons from different atoms from coming together must be defeated. This is accomplished by the addition of heat to create kinetic energy high enough to bring atomic nuclei close enough together for the short range strong nuclear force to become active, overwhelming the long range electrostatic or Coulomb force between protons.

The Sun has an internal temperature of 15.7 Million °C, which is exactly right for maintaining the fusion of hydrogen into helium. Its massive gravitational field provides an assist in keeping the burning plasma trapped so that continuous chain reactions can occur.

On Earth, we do not have a gravitational field strong enough to assist in this process. Furthermore, it is impossible to produce a container that can survive such high temperatures. However, there is a way. The addition of heat also brings hydrogen atoms above their ionization temperature. The electrons are stripped from their nuclei, and a plasma is created which contains positively charged nuclei, or ions, and negatively charged electrons. At this point, all the particles have a charge and therefore an electromagnetic field can be applied with a shape that keeps the plasma suspended without any need to touch container walls. Evidently, this must also occur in a vacuum so that air molecules are not implicated.

If this is already not enough of a technical challenge, there are other factors in play that require the plasma to be heated to much higher temperatures than exist even in the Sun's core! Primarily, we have to compensate for a relatively low density of the fuel which limits the probability of enough inter-nuclei collisions. So far, no experiment has yet been able to produce any fusion reaction that generates more energy than is input to create the plasma.

Here is the governing equation for a fusion reaction as planned in a design as described above, known as a Tokamak reactor.

$$^2_1H + {}^3_1H = {}^4_2He + \text{n} \quad \text{or;} \quad \text{D} + \text{T} + \text{He} + \text{n}$$

D stands for deuterium, otherwise known as heavy hydrogen. Unlike ordinary hydrogen, it contains a neutron. Deuterium can be harvested from water, which contains .031% "heavy" water. T stands for Tritium which contains 2 neutrons and is produced when the free neutrons in the equation above collides with container walls made of lithium. That tritium is then recycled back into the plasma. Using these two isotopes of hydrogen produces much more energy than the fusion of ordinary hydrogen atoms.

If .031% of all water on earth sounds small, it is enough to provide all the energy we would ever need for millions of years. One ton of deuterium contains the same amount of energy as 29 billion tons of coal! If we can get it to work, it literally provides unlimited clean energy.

How clean is it? The chamber walls that get bombarded by neutrons will become radioactive and will need to be treated properly. The scope of this problem is considerably less than with fission reactors. Additionally, there is no danger of a runaway reaction or meltdown as can be the case with nuclear fission. Any problem simply causes the reaction to stop.

The largest Tokamak reactor is currently under construction in France as part of an international research project called International Thermonuclear Experimental Reactor (ITER). As the name implies, it is purely experimental and is not intended to deliver electricity to the grid. It is the world's most important research facility for fusion power.

The plan is to achieve "First Plasma" in 2025. This stage will not produce any power, it will only demonstrate if the plasma can be generated and maintained. Then, in 2035 experiments will commence that will have the goal of generating fusion power using deuterium and tritium. There are many other projects going on in support of ITER. In 2020 the UK Atomic Energy Authority has already achieved "first plasma" on a smaller scale in a reactor called Mega Amp Spherical Tokamak (MAST).

Even if successful, it is doubtful whether this would lead to commercial electricity until about the year 2050. That may seem far off now, but it might be very welcome indeed when the time arrives.

Unlike the other GHG-mitigating technologies presented in this book, we must consider fusion as a medium risk of not ever being deployed. There are many undemonstrated technologies involved, and there is always the chance that the whole project might not be successfully realized. In spite of this concern, the potential is so spectacular that we must continue with the research until we have all the answers.

REVISITING WIND, SOLAR AND ELECTRIC VEHICLES

In Chapter 5 these options were developed in some detail. The conclusion was that 40% of electricity demand could be met by wind and solar, and 50% of the fleet of personal vehicles could be electric. Now we have to revisit these assumptions in light of the other options discussed in this chapter. How would they now fit in to the bigger picture?

Once we assume that the 3 nuclear options 5, 6 and 7 will become part of the story, then assigning wind and solar with 40% of electricity demand all the way out to year 2100 will no longer be sensible. Furthermore, some of the world has already recognized that using wind and solar to generate green hydrogen might be a better option. I will make the following assumptions: savings in emissions from wind and solar will initially follow the graph of chapter 5, which ramps up to 30% of electricity supply by 2040 and maintains it until 2100. However, starting in 2030 it will deliver half of its output to the electricity grid, and the other half to green hydrogen production.

For EV's, it seems reasonable to plan for ramping up to 50% penetration into the personal vehicle market by 2040 and maintain 50% all the way out to 2100. The benefit for EV over a gas-powered vehicle is initially 25% for the vehicle lifecycle up to year 2050. Then, we will assume that as electricity supply gets increasingly green, the benefit for EV will reach 100% by year 2100. Other cars and large trucks will be powered by hydrogen fuel cells once the technology and the green hydrogen is available, which I assume will start in 2040 for cars and 2050 for trucks. To be clear, that means 100% of trucks and 50% of passenger vehicles will be fuel cell vehicles by 2100.

From the Bossel paper (Ulf Bossel, 2001) I have extracted the following factors for hydrogen:

Amount of Hydrogen needed in a fuel cell vehicle to
provide performance equivalent to 2 kg of gasoline: .23753 kg
Electricity needed to generate 1 kg of useful
Hydrogen, including compression and transport: 57.34 KWh

A POSSIBLE PLAN OF ACTION

There are so many variables that it is clearly impossible to predict future outcomes with great accuracy. I will however attempt to show a plan that could reach a particular goal: the reduction of further temperature escalation by 2°C, to be accomplished by year 2100. The IPCC worst case scenario, which assumes no mitigation efforts, predicts a temperature escalation of 4°C by 2100. Therefore, this mitigation plan should limit escalation to 2°C. We will see how feasible this is after reporting on the total plan below.

This is not *the* plan; this is *a* plan. It is an *example* of a plan. It is full of assumptions, but I have explained what they are and how I got to them. Furthermore, future developments may render the plan progressively obsolete. However, it is a stake in the ground right now, and at the end of this chapter, we will take a step back to see if it is realistic.

SUMMARY OF ASSUMPTIONS

Wind and Solar: Ramp up to 30% of electricity demand by 2040, then maintain 30% out to 2100 as in Chapter 5. However, starting in 2030, send half of this energy to green hydrogen production, and the other half to the electricity grid.

Electric Vehicles: Ramp up to 50% market share by 2040 and maintain until 2100. Benefit in fuel savings will be 25% up until 2050, then ramp up to 100% in 2100 due to availability of green electricity.

Fuel Cell Cars: Start the introduction of fuel cell vehicles in 2040 and ramp up to 50% market share by 2100. Hydrogen will be delivered by electrolysis of water using electricity from wind, solar, and nuclear options Thorium fission, and Hydrogen fusion.

Fuel Cell Trucks: Start the introduction of fuel cell trucks in 2040 and ramp up to 100% market share by 2100. Hydrogen will be provided by the same sources as for cars.

AFOLU: Starting in 2025, ramp up to 5 Gtonnes of CO_2 savings in 2030, then maintain this as a constant until 2100.

Fossil Fuel Production: Starting in 2025, ramp up to 2030 where a savings of half of 2019 Methane and CO_2 emissions will be included. Maintain this as a constant until 2060. Then, ramp down to zero in 2100 assuming that there will be greatly reduced usage of fossil fuels.

Nuclear Power, Uranium: Ramp up from current 10.3% of electricity production to 22% by 2030. Maintain the resulting power delivery constant until 2050, then ramp down to zero by 2060. This plan will use most available reserves of Uranium.

Nuclear Power, Thorium: Start in 2030 and ramp up to about 25% demand by 2050. Maintain a constant power delivery until 2070, then ramp down to zero in 2080.

Nuclear Power, Fusion: Start in 2050 and ramp up until 2070 where it will provide all of the electricity demand except for the amounts provided by wind, solar and hydroelectric.

Industry: Green hydrogen will be the sole source of benefits provided to industry. Wind, solar, thorium and fusion will provide all of the electricity for electrolysis of water to generate hydrogen. If the Plan shows it to be possible, produce enough hydrogen to bring the total savings in temperature escalation to 2°C.

Energy Demand: Escalation of energy demand through the rest of this century is assumed to be as described in Chapter 5. It is a reduced rate as compared to recent history. Evidently, further reductions in demand would be beneficial.

Important Exchange Factors: Here is a summary of important factors that also appear elsewhere in this book.

When fossil fuels are removed from the electricity grid, savings are expected to be 733 kg CO2 per MWh of energy. This is a blended rate accounting for coal and natural gas applications.

Electricity needed to create 1 kg of green hydrogen is 57.3 KWh. This can be compared to the Heating value (HHV) of H_2 of 39.4 KWh, meaning that more energy is needed to create H_2 than is available for use. Then, burning of H2 in an industrial application is subjected to the same thermal efficiency debits as natural gas.

Energy output from hydrogen fuel cells for cars and trucks is taken from the Bossel paper, and the table of values is provided above.

The factor for converting emissions to temperature escalation is 1839.8 Gtonnes per °C as described in Chapter 5.

Here is a graphical representation of the results.

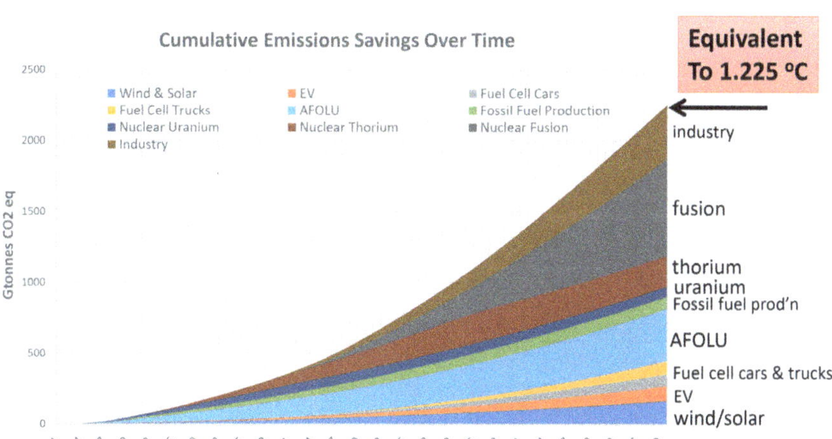

Before discussing this chart, a few clarifications are in order. These are cumulative emissions, meaning that the total at the end represents all of the savings from all sources. In this way, we can use the result to make a guess at how much temperature escalation is avoided. It is not exactly true, because the global temperature does not rise or fall instantaneously with emissions increase or reduction. However, I am using the exchange factor provided by the IPCC.

The cumulative nature of the chart makes it look like all the action is at the end. In fact, there are continuous emissions reduction throughout;

however, most of the strategies need time to develop because none of them have been started in earnest, with the exception of wind, solar and EV's. So, we have a lot of catching up to do.

The emissions savings are stacked one on top of another, so the total at the end represents all of the savings, and it translates into a reduction in temperature escalation of 1.225°C.

Another chart is necessary. This one shows the amount of green hydrogen generated in support of emissions reduction.

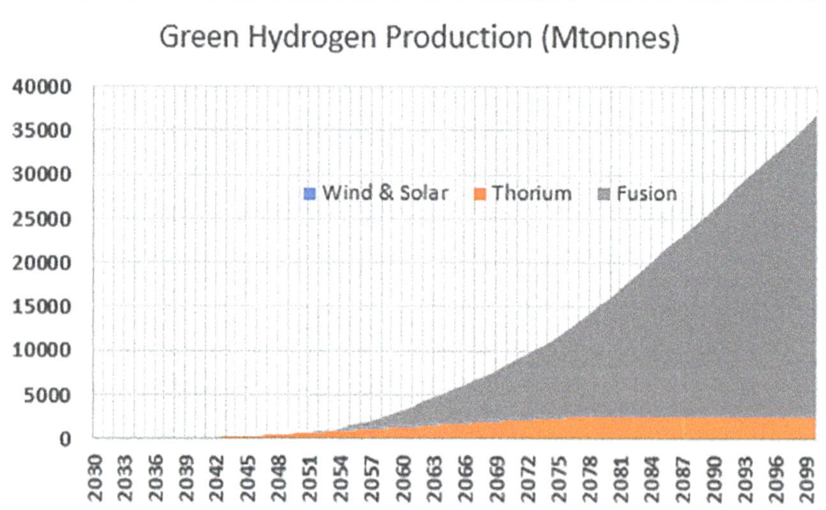

This plan is unable to reach the objective of removing 2°C of temperature escalation. We could continue to add more and more electricity from nuclear fusion during the second half of the century to create more benefit, but it feels unrealistic. The amount being produced in year 2100 by nuclear fusion is 90,000 TWh, which is about 3 times the total electricity production from all sources in 2019. It is also interesting that even with this aggressive plan, emissions have not gone to zero, even in 2100. Although all emissions were removed from electricity production and from transport, there are still emissions from industry and farming.

The implications of this are profound. Before elaborating, I would like to add one more calculation. This should be the last one in this book, I promise. If we somehow immediately cease all emissions of

CO_2 into the atmosphere, the world will continue to warm because at 420 ppm of CO_2, we have already shown that warming is occurring. I assume that we would have to gradually return to about 280 ppm in order to stop the warming because that was the level where global average temperature was stable for a very long time. The Earth will gradually absorb CO_2 by photosynthesis and absorption into soils and oceans until the concentration in the atmosphere drops to a balanced level. How long will this take?

I have found a few sources which claim that about half of our current emissions are absorbed by the Earth, and the other half remains in the atmosphere. Global CO2 emissions in 2019 were 33 Gtonnes. Therefore, I will assume that 16.5 Gtonnes a year can be absorbed by the Earth. If we hold this value constant, here is the calculation.

Weight of the Atmosphere Gtonnes	5.15E+06
Mass of CO_2 in air at 420 ppm	2.16E+03
Target Mass at 280 ppm	1.44E+03
Amount to be Removed	7.21E+02
Reduction per Year	16.5
No of Years to Meet Target	43.7

It will take 44 years to reach stability even without any new emissions.

It seems we have passed the point of no return, and the world will continue to warm, barring a startling new invention or process that can dig us out of this hole. As already discussed in Chapter 5, the exchange factor between emissions reduction and temperature reduction is extremely high. This is because the Earth has a roughly constant ability to absorb CO_2, so the portion remaining in the atmosphere increases each year unless emissions are reduced to the level that can be absorbed. This has been called carbon neutrality. Then, the drawdown of temperature can begin, but that will be a slow process. By then, I expect we will be living in a much warmer world and we will have to deal with it.

I am myself quite startled with this conclusion. I was thinking that all the technology that I was considering could reverse the trend. It seems not to be the case.

Does this mean we should not follow a plan like the one I have presented? Certainly not. As a minimum, carbon neutrality is eventually reached. I have estimated emissions in year 2100 using the following assumptions:

> Industrial emissions that could not be removed with the level of production of green hydrogen the Plan supplies (8.9 Gtonnes).
>
> Emissions from buildings which I have unfortunately ignored in this book. However, I will assume 2100 value to be equal to 2019 value, meaning that efficiency gains eliminate any ramp up of energy use in buildings (2.6 Gtonnes).
>
> Emissions from aircraft which seems to me unlikely to be mitigated in any way. Increase in emissions are assumed to follow the rate of ramp-up of energy demand (1.4 Gtonnes).

When I do this sum, I arrive at remaining emissions of 12.9 Gtonnes of CO_2. As we established previously, the Earth can naturally absorb 16.5 Gtonnes per year, so we have exceeded the carbon neutrality objective. In fact, there is some room to also offset methane and nitrous oxide emissions from agriculture.

I need to include here a reminder from Chapter 4 that fossil fuels are not in infinite supply. In this extremely aggressive plan, they are still being used, although at a reduced rate, all the way out to year 2100. Will they even be available? The answer is probably yes, but we will be at the point where extraction from less bountiful sources will be exceedingly difficult and expensive. I repeat that we need to consider these fuels as precious resources. This study shows how incredibly powerful they are. They are a compact form of energy developed over many millions of years by storing the sun's energy in carbon-based life forms. Now we will have to find other ways to create such energy, and it will be very expensive.

THE GEOPOLITICAL LANDSCAPE - REDUX

Now that I have gotten to the end, here is a sprinkling of thoughts and observations in the geopolitical world.

I am bewildered by all the promises to achieve carbon neutrality by 2050. Countries that make that promise do not present any detailed plan like the one I have shown here, so I suspect that promises will be broken.

A group called the Energy Transitions Committee issued a press release stating that between $1 and $2 Trillion annual investment will be needed to get us to carbon neutral by 2050. The plan would include quadrupling of electricity output to counteract fossil fuels, and the use of green hydrogen. There is no mention of any technology to reach that goal except for wind and solar. It happens that the plan I have presented reached that level of electricity generation, but only in the year 2100, and it uses mostly nuclear fusion which only kicks off in 2050. Perhaps they have another plan. I hope so.

Every debate talks about renewables as the savior, meaning wind and solar. On my charts they can hardly be seen they are so small in their contribution. A TV commercial for Amazon states that as a company they will be carbon neutral by 2050. Incredibly, they admit that they do not yet know how to get there. The usual images of windmills and electric cars appear.

Nuclear power is still seen as an unacceptable choice by most countries. The Green Party in Germany has succeeded not only in stopping new development of nuclear power, but also in shutting down existing ones that are perfectly productive. Meanwhile, new coal power plants appear.

Every country has carbon emission limits. No-one seems to be aware that all of the commitments add up to zero when China's expected increase in emissions is considered.

If I go ahead and assume that my calculations reported in this book are in the right ballpark, then there is a widespread unfortunate lack of awareness of the nature of the beast we need to tame. I am not sure why this should be so. There are certainly experts in the field who are much more qualified than me to create these projections. We certainly hear of doom and gloom projections on a qualitative basis, but someone should step up and correct misconceptions about the hard facts.

PREPARING FOR CLIMATE CHANGE

In this book I have already concluded that current plans to mitigate Climate Change are inadequate. Furthermore, many items listed above that might have powerful positive impacts are still quite far away from yielding results, even if world leaders decide to adopt them. Therefore, I need to add here a part of the plan that prepares for the effects of Climate Change. This may be unpopular because it could be perceived as giving up on the war and conceding defeat. I have learned in this journey that the world is going to continue to warm, and the extent to which we can mitigate it will be limited in the short term but, hopefully, profound in the long term.

As I stated elsewhere in this book, Climate Change is not a monolithic issue. Everything is tied together in some way. For example, Climate Change will put stresses on wildlife, but the pressures we are already putting on many animal species are probably greater in magnitude. Deforestation is harmful in more ways than just loss of carbon storage. Overpopulation along riverbanks in Asia and Africa, and in California desert regions are already foolhardy, even before the worst Climate Change effects are realized. While we worry about inundation of the Atlantic Ocean onto Miami Beach, it has already eroded to such an extent that we can conclude that every grain of sand on it today has been placed there by humans. Long before any significant sea level rise occurs, the beach would disappear completely, with all the sand swept back into the ocean, if not for human efforts to maintain it.

Perhaps Climate Change can be the flag that we fly when we try to correct all of these issues, and many more. I believe we should go ahead and assume that temperatures will continue to rise significantly, perhaps to a level of +3°C, and then devise strategies to deal with it. The outcomes of proper planning can only be positive. Here is a partial list of some of the things we could do.

1. Re-imagine our coastlines, assuming a sea level rise of 1 meter or more. This could mean a moratorium on building any new structures in the danger zone, and a gradual movement of

existing structures and populations. There is plenty of time to do these things in an orderly and cost-effective manner.

2. Reinforce riverbanks to effect containment of flood waters. Move people as required, again in a gradual way.

3. Reduce population expansion into zones which are known to be too hot and dry.

4. Place a moratorium on deforestation in tropical rain forests and northern boreal forests.

5. Identify plant and animal species that are at risk and devise long term plans for their survival.

In all of these cases, there will be supplementary benefits to be had. For example, while shoring up river banks, we could take the opportunity to create water reservoirs for impoverished communities that do not currently have adequate access to clean potable water, and we could eliminate the infusion of waste plastics into those very same rivers.

In developing nations, it may be necessary to extract funding and expertise from wealthier nations. So be it.

If we prepare for Climate Change, it will make us all feel like responsible citizens of the world, and perhaps take away some of the angst that people feel about it.

CONCLUSIONS FROM THIS CHAPTER

Many possible sources of emissions reductions have been evaluated quantitatively and the result shows a collective reduction in temperature escalation of around 1.2 °C by year 2100. The target of 2°C is not met in spite of large aggressive investments in new energy infrastructure. However, carbon neutrality is reached in 2100 if this plan were to be carried out.

The plan includes large new development in nuclear power including conventional Uranium-based fission reactors, Thorium reactors and nuclear fusion technology. The latter can only be expected to start delivery in 2050 and only if many undemonstrated technologies become successful.

Solutions currently in the public eye, primarily wind, solar and electric vehicles, are feasible but cannot make large contributions to emissions reductions.

The generation of hydrogen with wind, solar and nuclear power is proposed to power fuel cell vehicles and the needs of various industrial processes. The needs of industry are only met by about half in spite of large new electricity generated to create hydrogen by electrolysis.

Fossil fuels gradually diminish in importance but remain a staple all the way out to year 2100. It is uncertain whether there will be sufficient supply.

An estimate of 5 Gtonnes per year is used for emissions reduction from farming and forestry (AFOLU). This value is on the low side of estimates provided by IPCC documents. Considerable interventions by governments will be needed to deliver these savings.

Given that warming is expected to continue for the rest of this century, it is recommended to make appropriate preparations for mitigation of the effects.

There seems to be poor understanding of the future by decision makers and the media. It is expected that switching to renewables such as wind and solar will solve most problems and carbon neutrality can be achieved by 2050. Based on the studies completed in this chapter, these are highly unlikely scenarios.

CHAPTER 8

CONCLUSIONS FROM THIS BOOK

WARMING OF THE PLANET IS validated to be a real phenomenon, and it is caused by greenhouse gases emitted from human activity.

Effects of warming in the current world are exaggerated in the media and even by some scientists. For example, there is no evidence that tropical storms are any worse than they have always been. The effects so far validated to be real are melting of sea ice and glaciers from the poles, worsening of some arid environments, and stress on certain tropical species such as corals.

Effects of warming in the future are likely to be quite drastic if we continue without large reductions in emissions. Hot and dry areas will get hotter, and some might become unlivable. There will be increased rainfall causing flooding of riverbanks, especially in developing countries. Sea levels will rise enough to cause the necessity of moving large populations. Many animal species will need to move to cooler areas. Humans in all areas will feel the heat, especially on peak summer days.

Fossil fuels may run out before the worst effects of warming are felt. Humans will need to extract them from increasingly difficult processes and locations. Short term drawdown of fossil fuels would have much more dramatic effect than is being claimed, including negative effects on food security. These fuels should still be treated as precious resources if we plan to maintain comfortable existence for more than 7 billion people.

Mitigation of emissions that are currently in the public eye are useful, but they will not make a large contribution. Wind power, solar power, and electric vehicles together cannot produce more than about a quarter of a degree Celsius of impact on the warming world.

Governments around the world repeatedly make emission reduction promises that they do not keep. World leaders have short term political objectives while long term strategy is required. The European Union, led by Germany and the UK, has been an exception and has made proven progress. The Paris Accord is a promising initiative in the sense that all countries have made commitments. However, all of the commitments combined add up to zero savings by 2030, primarily because China's emissions will greatly increase.

A detailed plan has been laid out in this book that creates a best-case scenario where very aggressive strategies are used to bring down emissions. These include nuclear power in three forms: conventional fission with Uranium235, generation of fissionable Uranium233 using Thorium as a breeder, and nuclear fusion of Hydrogen into Helium. The latter technology would only be available commercially by around 2050, and there is some risk that it may not work at all. New electricity sources are used to generate hydrogen for industrial use and fuel cell vehicles.

The plan fails in its target of reducing temperature escalation by 2°C. It reaches only 1.2°C in spite of large new investments. However, carbon neutrality is reached in year 2100. This is the condition where emissions equal the Earth's ability to absorb the carbon generated.

It is recommended to start planning for a significant level of warming because it appears from the development of this plan that we cannot fully compensate for the task ahead, barring any new discoveries that may come along.

The awareness of our fate seems to be exceptionally low among world leaders and the media. There is an impression that "switching to renewables" will save the day. This is simply not so. Furthermore, commitments to carbon neutrality by 2050 do not seem to be accompanied by any details plans on how to get there. I hope such plans are found. I was not able to do so.

AFTERWORD

IN SPITE OF THE SOMEWHAT gloomy conclusions, I feel there is some success in the journey I have made in the writing of this book. I feel somewhat enlightened by the findings and I hope that some of them can make it out into the global conversation.

I feel cautiously optimistic about the future because there is tremendous power in knowing your enemy. Attaining that knowledge was indeed the objective of this book. I only hope it has been transmitted in a useful way.

Thanks for reading.

BIBLIOGRAPHY

Anna L. Jacobsen, R. B. (2018). Extensive drought-associated plant mortality as an agent of type-conversion in chaparral shrublands. *New Phytologist.*

Arti Bhatia, N. J. (2012). *1© 2013 Society of Chemical Industry and John Wiley & Sons, Ltd | Greenhouse Gas Sci Technol. 1–16 (2013); DOI: 10.1002/ghgCorrespondence to: A. Bhatia, Centre for Environment Science and Climate Resilient Agriculture, Indian Agricultural Research Insti.* New Delhi, India: Centre for Environment Science and Climate Resilient Agriculture, Indian Agricultural Research Institute.

Bernhard Bereiter, S. E.-A. (2015). Revision of the EPICA Dome C CO2 record from 800 to 600 kyr before present. *Geophysical Research.*

BROMWICH, J. P. (2014). New Reconstruction of Antarctic Near-Surface Temperatures: Multidecadal Trends and Reliability of Global Reanalyses. *Journal of Climate.*

Collins, M. R.-L. (2013). *Long-term Climate Change: Projections, Commitments and Irreversibility.* Cambridge, United Kingdom and New York, NY: Cambridge University Press.

Curtis A. Deutsch, J. J. (2008). Impacts of climate warming on terrestrial ectotherms across latitude. *Proceedings of the National Academy of Sciences of the United States of America (PNAS).*

D. Lew, G. B. (2012). *Impacts of Wind and Solar on Fossil-Fueled Generators .* San Diego, Cal USA: NREL.

David P. Schneider, E. J. (2006). Antarctic Temperatures Over the Past Two Centuries from Ice Cores. *Geographical Research Letters.*

DOLAN, S. L. (2012). Life Cycle Greenhouse Gas Emissions of Utility-Scale Wind Power. *Journal of Industrial Ecology*.

Fetterer, F. (2016). *Guest post: Piecing together the Arctic's sea ice history back to 1850*. Carbon Brief: https://www.carbonbrief.org/guest-post-piecing-together-arctic-sea-ice-history-1850.

Field, C. V. (2014). *Climate Change 2014: Impacts, Adaptation, and Vulnerability. Part A: Global and Sectoral Aspects*. Cambridge, United Kingdom and New York, NY: Cambridge University Press.

Fischedick M., J. R.-A.-P. (2014). *Climate Change 2014: Mitigation of Climate Change. Contribution of Working Group III to the Fifth Assessment Report of the Intergovernmental Panel on Climate Change*. Cambridge, UK: Cambridge University Press.

Fubao Sun, M. L. (2018). Rainfall statistics, stationarity, and climate change. *Proceedings of the National Academy of Sciences 115(10):201705349*.

G.H. Miller, J. B.-G. (2010). Temperature and precipitation history of the Arctic. *Quaternary Science Reviews*.

Garcia, B. C. (2018). Climate-driven declines in arthropod abundance restructure a rainforest food web. *Proceedings of the National Academy of Sciences of the United States of America (PNAS)*.

Gerald R. North, F. B. (2006). *Surface temperature reconstructions for the last 2,000 years*. National Research Council (U.S.). Committee on Surface Temperature Reconstructions for the Last 2,000 Years.

Harde, H. (2017). Radiation Transfer Calculations and Assessment of Global Warming by CO2. *International Journal of Atmospheric Sciences*.

Haseli, N. S. (2019). A Critical Review of CO2 Capture Technologies and Prospects for Clean Power Generation. *Molecular Diversity Preservation International*.

Huihui Feng, M. Z. (2015). Global land moisture trends: Drier in dry and wetter in wet over land. *Scientific Reports*.

J. M. Lough, K. D. (2018). Increasing thermal stress for tropical coral reefs: 1871–2017. *Scientific Reports*.

Jérémie Mouginot, E. R. (2018). *Forty-six years of Greenland Ice Sheet mass balance*. PNAS.

John A. Church (Australia), P. U. (2014). *Sea Level Change*. Cambridge, United Kingdom and New York, NY: Cambridge University Press.

Jon E Keeley, A. S. (2018). Historical patterns of wildfire ignition sources in California ecosystems. *International Journal of Wildland Fire.*

Joris van Loenhout, R. B. (2020). *Human Cost of Disasters.* Brussels: UN Office for Disaster Risk Reduction.

Julie Arblaster (Australia), J.-L. D. (2013). *Long-term Climate Change: Projections, Commitments and Irreversibility.* Cambridge, United Kingdom and New York, NY: Cambridge University Press, Cambridge, United Kingdom and New York, NY, USA.

K. R. N. Anthony, D. I.-P.-G. (2008). Ocean acidification causes bleaching and productivity loss in coral reef builders. *PNAS.*

Keith, A. S. (2013). Are global wind power resource estimates overstated? *Environmental Research Letters.*

Keith, L. M. (2018). Observation–based solar and wind power capacity factors and power densities. *Environmental Research Letters.*

Kopp, G. (2016). *Magnitudes and Timescales of Total Solar Irradiance Variability.* University of Colorado, Laboratory for Atmospheric and Space Physics.

Lee M. Miller, D. W. (2018). Climatic Impacts of Wind Power. *Joule.*

Li Ren, P. A. (2013). Global precipitation trends in 1900–2005 from a reconstruction and coupled model simulations. *Journal of Geophysical Research: Atmospheres.*

Li-Qing Jiang, B. R. (2019). Surface ocean pH and buffer capacity: past, present and future. *Nature/Scientific Reports.*

Lövseth, R. N. (1996). *CARBON-14 MEASUREMENTS IN ATMOSPHERIC CO2 FROM NORTHERN AND SOUTHERN HEMISPHERE SITES, 1962-1993.* Trondheim, Norway: Radiological Dating Laboratory, The Norwegian Institute of Technology .

Lutsey, D. H. (2018). *Effects of battery manufacturing on electric vehicle life-cycle greenhouse gas emissions.* International Council on Clean Transportation.

Megha Maheshwari, R. K. (2013). An Investigation of the Southern Ocean Surface Temperature Variability Using Long-Term Optimum Interpolation SST Data. *Hindawi Publishing Corporation, ISRN Oceanography.*

Melillo, J. M. (2014). *Climate Change Impacts in the United States: The Third National Climate Assessment.* Washington, USA: U.S. Global Change Research Program.

Metz, B. D. (2005). *IPCC, 2005: IPCC Special Report on Carbon Dioxide Capture and Storage.* Cambridge, UK: Cambridge University Press.

Miguel Angel Gonzalez-Salazara, T. K. (2018). Review of the operational flexibility and emissions of gas- and coal-fired power plants in a future with growing renewables. *Renewable and Sustainable Energy Reviews.*

Mohammad Reza, A. J. (2020). A century of observations reveals increasing likelihood of continental-scale compound dry-hot extremes. *Scienec Advances.*

Nic Newman with Richard Fletcher, A. S. (2020). *Reuters Institute Digital News Report 2020.* Oxford, UK: Reuters Institute.

Nikolaos Skliris, J. D. (2016). Global water cycle amplifying at less than the Clausius-Clapeyron rate. *Scientific Reports.*

O. Liik, R. O. (2003). Estimation of real emissions reduction caused by wind generation. *International Energy Workshop, 24-26 June 2003, IIASA.* Laxenburg, Austria.

P. Bousquet, P. C. (2006). Contribution of anthropogenic and natural sources to atmospheric methane variability. *Nature/Letters.*

Paire, C. (. (n.d.). Facts About the Savings of Fossil Fuel by Windturbines in the Netherlands .

Peter Greve, B. O. (2014). Global assessment of trends in wetting and drying over land. *Nature Geoscience .*

Peter Soroye, T. N. (2020). Climate change contributes to widespread declines among bumble bees across continents. *Science.*

Pirani, S. (2018). *Burning Up.* London, England: Pluto Press.

R Camilla Thomson, G. P. (2015). *Life Cycle Costs and Carbon Emissions of Onshore Wind Power.* Edinburgh, Scotland: ClimateXChange.

Reconstructions for the Last 2,000 Years Surface temperature reconstructions for the last 2,000 years. (n.d.).

Report, B. (2020). *Statistical Review of World Energy 2020 69th Edition.* London, England: BP.

Robert E. Kopp, R. M. (2017). *Evolving Understanding of Antarctic Ice-Sheet Physics and Ambiguity in Probabilistic Sea-Level Projections.* Earth's Future.

Roser, H. R. (2017). *"CO$_2$ and Greenhouse Gas Emissions". Published online at* OurWorldInData.org. *Retrieved from:* 'https://ourworldindata.org/co2-and-other-greenhouse-gas-emissions' *[Online Resource].* Our World in Data.

S. C. Pryor, R. J. (2020). 20% of US electricity from wind will have limited impacts on system efciency and regional climate. *Scientific Reports.*

Saunois, M. (2020). The Global Methane Budget 2000–2017. *Earth System Science Data.*

Smith P., M. B. (2014). *Agriculture, Forestry and Other Land Use (AFOLU).* Cambridge, UK: Cambridge University Press.

T, M. T. (2016). *Climate change and the Great Barrier Reef, Policy Information Brief 1.* Gold Coast: National Climate CHange Adaptation Research Facility (NCCARF).

Udo, F. (n.d.). *Wind energy in the Irish power system.* National Wind Watch.

Ulf Bossel, B. E. (2001). *Energy and the Hydrogen Economy.*

Victor, D. D.-H. (2014). *Introductory Chapter. In: Climate Change 2014: Mitigation of Climate Change. Contribution of Working Group III to the Fifth.* Cambridge, United Kingdom and New York, NY, USA: Cambridge University Press.

Wang, T. (2020). *U.S. annual precipitation volume 1900-2019.*

Zyss, N. J. (2019). Improvements in the GISTEMP Uncertainty Model. *JGR Atmospheres.*

ACKNOWLEDGMENTS

MANY THANKS TO 3 DEAR friends (and scientists in their own right) for their careful review of my preliminary manuscript and their useful inputs: Nicolas Grivas, Kapila Jain and Douglas MacCaul.